Handbook of Maize Production

Handbook of Maize Production

Hazem Shawky Fouda

Delve
PUBLISHING
www.delvepublishing.com

Handbook of Maize Production

Hazem Shawky Fouda

Delve Publishing
224 Shoreacres Road
Burlington, ON L7L 2H2
Canada
www.delvepublishing.com

Email: orders@arclereducation.com

ISBN: 978-1-77469-125-0 (Hardcover)

ABOUT THE AUTHOR

Hazem Shawky Fouda has a PhD. In Agriculture Sciences, obtained his PhD. from the Faculty of Agriculture, Alexandria University, 2008, MSc. In Agriculture Sciences from the Faculty of Agriculture, Alexandria University in 2004, Post-Grade Diploma in Cotton, 2001, BSc. in Agriculture Sciences, from the Faculty of Agriculture, Alexandria University, 1997, He worked in Cotton Arbitration & Testing General Organization (CATGO) since 1999 till 2018. Was working in the International Cotton Training Center (ICTC) – Cotton Arbitration & Testing General Organization (CATGO) from 2000 till 2015, as a Lecturer & Classer's Trainer for Egyptian and foreigner classers, technicians, ginners, spinners & traders in all cotton aspects starting from picking seed cotton till preparing bales for export including other cotton aspects and training them practically on classing cotton. Besides that he was an editor and active member in the Research & Translation Committee, participating in issuing weekly, monthly and annually issues about the international & local cotton market including price trends and direction, recent developments & researches concerning cotton production, protection, harvesting, ginning, fiber testing, spinning & weaving since its foundation in 2000 till 2014 and since 2015 till 2018 he worked as an inspector, from 2018 till present works as a consultant.

TABLE OF CONTENTS

List of Figures .... *xi*

List of Tables .... *xiii*

List of Abbreviations .... *xv*

Acknowledgement .... *xvii*

Preface .... *xix*

Chapter 1 History and Origin of Maize 1

1.1. Introduction 2

1.2. The Origin of Maize 7

1.3. The Natural History of Maize 12

1.4. Maize in the New World 15

1.5. Maize in the Old World 17

1.6. The African Connection 18

1.7. The Future of Maize 22

1.8. Conclusion 23

References 24

Chapter 2 Land Preparation Practices for Maize Production 27

2.1. Introduction 28

2.2. Methods Involving no Tillage or Seedbed Shaping 29

2.3. Animal versus Tractor Power: Some Considerations for the Small Farmer 35

2.4. Animal-Drawn Disk Harrow 38

2.5. Beans Grown on Raised Beds 39

2.6. Gear for Seedbed Shaping 41

2.7. Step-by-Step Guide to Planting Maize 43

2.8. Land Preparation Methods for Maize Production 46

2.9. Conventional Till Flat Planting 48

2.10. Impacts of Land Preparation 49

2.11. Conclusion 51
References 52

Chapter 3 Climate and Soil Requirement of Maize Farming 53
3.1. Introduction 54
3.2. Agroclimatology of the Crop 56
3.3. Climatic Requirements 58
3.4. Weather Stress in Maize Cultivation 61
3.5. Soil: The Foundation of Agriculture 64
3.6. Soil Degradation and Crop Production 69
3.7. Sustainable Soil Management 70
3.8. Soils for Maize 71
3.9. Soil Requirements for Growing Corn 72
3.10. Conclusion 75
References 77

Chapter 4 Maize Seed Production 79
4.1. Introduction 80
4.2. Producing Quality Seeds Means Quality Yields 83
4.3. Creating Seed-Growers' Cooperatives 84
4.4 Quality Seed Use Doubles 85
4.5. Maize Hybrid Seed Production 86
4.6. Maize Hybrid 87
4.7. Seed Classes 89
4.8. Significant Considerations for Hybrid Seed Production 94
4.9. Conclusion 98
References 100

Chapter 5 Growth and Development of Maize 101
5.1 Introduction 102
5.2 The Maize Plant 103
5.3. Corn Growth & Development 107
5.4. Germination and Emergence 109
5.5. Reproductive Development 116
5.6. Conclusion 118
References 119

Chapter 6 **Weed Management in Maize** 121

6.1. Introduction 122

6.2. Weeds 125

6.3. Major Weed Species in Association with Maize 128

6.4. The Critical Period of Crop-Weed Competition 130

6.5. Weed Management Program in Maize 133

6.6. Effect of Pre-Sowing Herbicides on Weed Control of Maize 138

6.7. Impact of Pre-Rise Herbicides on Weed Control in Maize 139

6.8. Effect of Post-Emergence Herbicides on Weed Control in Maize 140

6.9. Integrated Weed Management in Corn 141

6.10. Conclusion 142

References 143

Chapter 7 **Agronomic Practices of Maize** 145

7.1. Introduction 146

7.2. Concepts And Guiding Principles of Good Agricultural Practices (GAP) 148

7.3. The GAP Approach and Concepts 150

7.4. Integrated Pest Management (IPM) 152

7.5. Integrated Crop Management (ICM) 153

7.6. Tips on Best Practice Maize Production 154

7.7. Maize Agronomic Principals 157

7.8. Conclusion 162

References 164

Chapter 8 **Integrated Pest Management in Maize** 165

8.1. Introduction 166

8.2. Diseases Pest Management Approach 168

8.3. Management Practices for Insect Pests 170

8.4. Pest Management 172

8.5. General Guidelines for Management of Resistance 180

8.6. Safety Parameters in Pesticides Usage 181

8.7. Insect Pest Management In Maize 182

8.8. Conclusion 189

References 190

Chapter 9 **Advanced Maize Production Technologies ... 191**

9.1. Introduction ... 192

9.2. Improved Technologies for Higher Maize Production ... 196

9.3. Improved Production Technologies ... 198

9.4. Other New Technologies Available Today ... 209

9.5. Production technology of maize ... 213

9.6. Conclusion ... 215

References ... 217

Chapter 10 **Challenges in Maize Production ... 219**

10.1. Introduction ... 220

10.2. Problems and Issues for White Maize Production ... 226

10.3. Difficulties and Opportunities in Maize Production ... 228

10.4. Fertilization and Liming ... 230

10.5. Crop Rotation ... 232

10.6. Development and Cultivation ... 232

10.7. Problem Descriptions ... 232

10.8. Management Factors ... 236

10.9. Maize in the Developing World: Trends, Challenges, and Opportunities ... 240

10.10. Conclusion ... 242

References ... 244

Index ... 245

LIST OF FIGURES

Figure 1.1. History and origin of maize
Figure 1.2. Structure of sucrose
Figure 1.3. Taxonomic rank graph
Figure 1.4. An illustration of maize – teosinte
Figure 1.5. An illustration of archaeologists of Hazara University
Figure 2.1. Methods involving no tillage or seedbed shaping
Figure 2.2. Animal versus tractor power: some considerations for the small farmer
Figure 2.3. Various shapes of seedbed
Figure 3.1. Climatic requirements for maize production
Figure 3.2. Weather stress in maize production
Figure 3.3. Soil: foundation of agriculture
Figure 3.4. Soil requirements for growing corn
Figure 4.1. Production of maize seed
Figure 4.2. Scientist examining seeds in fields
Figure 4.3. Image of small - scale grain cooperative farmer
Figure 4.4. Image of hybrid seed production
Figure 5.1. Image of coleoptile
Figure 5.2. Image of corn leaf
Figure 5.3. Image of stalk of corn
Figure 5.4. Image of corn tassel
Figure 5.5. Image of seed germination of maize
Figure 6.1. An illustration of weed plant
Figure 6.2. Weed are found to be harmful to crops, cattle and humans as well
Figure 6.3. There are various benefits of weed in medical science
Figure 6.4. Major weed species in association with maize
Figure 6.5. An illustration of cannabis plant
Figure 6.6. Weed control in zero-till maize by chemical measures
Figure 6.7. Impact of pre-rise herbicides on weed control in maize
Figure 7.1. Sustainable agriculture

Figure 7.2. Integrated Pest Management (IPM) Cycle
Figure 7.3. Best practices for maize production
Figure 7.4. Openings for non-mechanized conservation agriculture
Figure 7.5. Maize agronomic principals
Figure 7.6. Worsen climate impact on maize production
Figure 7.7. Effect of nutrition on forage maize yield
Figure 7.5. Harvesting of maize
Figure 7.8. Optimum harvest stage
Figure 8.1. Dynamic sketch of IPM system
Figure 8.2. Management practices for insect pests
Figure 8.3. Diversification by genetic management
Figure 8.4. Stem borers hidden in maize tassel
Figure 8.5. Soil bugs
Figure 8.6. Moles by bug
Figure 9.1. Climate Smart crop production practices
Figure 9.2. Nutrient management
Figure 9.3. Maize stem borer
Figure 9.4. Solar tech in maize production
Figure 9.5. Data management technologies
Figure 10.1. Challenges in maize production
Figure 10.2. Problems and issues for white maize production
Figure 10.3. Difficulties and opportunities in maize production
Figure 10.4. Fertilization and liming
Figure 10.5. Climatic factors and soil conditions
Figure 10.6. Sun and temperature impacts on maize fields
Figure 10.7. High temperature stress and drought probably disturbs the maize yielding

LIST OF TABLES

Table 3.1. Essential plant nutrient elements and their primary form utilized by plants
Table 7.1. Treatment details

LIST OF ABBREVIATIONS

AB	Accreditation Body
AESA	Agro-Ecosystem Analysis
AN	Ammonium Nitrate
BT	Bacillus Thuringiensis
CA	Conservation Agriculture
CB	Certificate Body
DM	Dry Matter
EGP	Eastern Gangetic Plains
ESA	Eastern and Southern Africa
FAW	Fall Armyworm
FTL	Financial Limit Level
FYM	Farm Yard Manure
GAP	Good Agricultural Practices
GVB	Green Vegetable Bugs
HSF	High Soil Fertility
IAP	Coordinated Agronomic Practice
ICAR	Indian Council of Agricultural Research
ICM	Integrated Crop Management
IFRI	International Food Research Policy Institute
IMPACT	International Model for Policy Analysis of Agricultural Commodities and Trade
IPM	Integrated Pest Management
IWM	Incorporated Weed Management
LAI	Leaf Area Index
LSF	Low Soil Fertility
MAP	Mono Ammonium Phosphate
MSV	Maize Streak Virus
NUE	Nitrogen Use Productivity
RP	Ridge Planting

SO	Scheme Owner
VR	Variable Rate
WCA	Western and Central Africa

ACKNOWLEDGEMENT

First and foremost, I feel always indebted to **ALLAH**, the most beneficent and merciful.

Words stand short when they came to express my gratefulness to my late parents **Shawky Fouda** and **Fatma El-Hussainy**, my uncle **Ahmed Fouda**, my wife **Heba Sadek**, my son **Mohammed Fouda**, my daughter **Raghad Fouda,**

And a special thanks to my sister **Riham Fouda,** relatives and friends for their support, encouragement, care and patience and for their love.

PREFACE

This book takes the readers through several different stages and concepts of maize production, its history, land preparation practices, climate and soil requirement for its production, maize seed production, growth and development of maize. This book sheds light on weed management in maize, agronomic practices of maize, integrated pest management in maize, different technologies implemented, and the challenges in its production.

The first chapter stresses on the basic overview of the history and origin of maize, taxonomy, maize in the light of the new world. This chapter will also emphasize the natural history of the maize, and its concept in the old world. The African connection and the future of maize.

The second chapter takes the readers through the concepts of land preparation practices for the production of maize, the different methods involved. This chapter will provide highlights on the animal versus tractor power, animal drawn disk harrow, beans grown on raised beds, conventional tillage, flat planting as well as the land preparation.

Then, the third chapter explains the climate and soil requirement of maize farming, agroclimatology of the crop. It also explains the climatic requirements, weather stress in maize cultivation, and soil as the foundation of agriculture. This chapter also sheds light on the sustainable soil management and the requirements of soil for growing corns.

The fourth chapter introduces the readers to the aspect of maize seed production, producing quality seeds, creating of the seed-growers cooperatives. This chapter also explains the concept of quality seed use doubles, maize hybrid seed production, and significant considerations for hybrid seed production.

The fifth chapter throws light on the growth as well as the development of maize, different kinds of maize plant. This chapter contains different approaches to the corn growth and development, germination and emergence, vegetative growth and the reproductive development.

The sixth chapter takes the readers through concept of weed management in maize, characteristics of weed. The readers are then told about the critical period of crop-weed competition, weed management programs in maize such as social control, flame weeding etc. It then talks about the impact of pre-rise herbicides on weed control in maize.

The seventh chapter explains the agronomic practices of maize, the concepts and guiding principles of good agriculture. This chapter also emphasizes on the GAP approach and concepts, and the maize agronomic principles

The eighth chapter of this book sheds lights on the integrated pest management in maize and the diseases pest management approach. This chapter also mentions the management practices for insect pests, pest management, different types of guidelines for management, as well as the safety parameters in pesticides usage.

The ninth chapter of this books gives insight to the reader on the advanced maize production technologies, the improved technologies for higher maize production, nutrient management. It also illustrates the insect and pest management, new technologies available in recent years, and the data management technologies.

The tenth chapter of this book sheds light on the challenges, the problems, issues as well as the opportunities in maize production. It talks about the fertilization and liming, problem descriptions like the climatic factors, management factors etc. and it also defines maize in the context of the developing world.

This book has been designed to suit the knowledge and pursuit of the researcher and scholars and to empower them with various concepts of the production of maize and its different dimensions like the history, requirements, challenges, its development and the management practices. I hope that the readers find the book explanatory and insightful and that this book is referred by the scholars across various fields.

Chapter 1

History and Origin of Maize

CONTENTS

1.1. Introduction 2
1.2. The Origin of Maize 7
1.3. The Natural History of Maize 12
1.4. Maize in the New World 15
1.5. Maize in the Old World 17
1.6. The African Connection 18
1.7. The Future of Maize 22
1.8. Conclusion 23
References 24

This chapter deals with the history and origin of maize and its historical argument as well as the modern argument. It sheds light on the natural history of maize, its significance. It also talks about the ethnobotany of maize and describes maize in the light of the quintessential maize plant.

It also defines maize in the context of a new world along with explaining its historical backdrops as well as it illustrates maize in the old world which constitutes several countries such as China and Southeast Asia. It gives insight to the reader on the concept of old-world dispersals. The social, financial and political effects.

It goes on putting emphasis on the African connection, the future of maize that elucidates how the maize is prepared into a bewildering cluster of consumer products.

1.1. INTRODUCTION

Corn, (Zea mays), additionally called Indian corn or maize, oat plant of the grass family (Poaceae) and its edible grain. The domesticated yield started in the Americas and is possibly the most broadly circulated of the world's food crops.

Corn is utilized as animals feed, as human food, as biofuel, and as raw material in industry. In the United States the vivid variegated strains known as Indian corn are customarily utilized in fall harvest enhancements.

Corn was first domesticated by local people groups in Mexico around 10,000 years prior. Local Americans encouraged European pilgrims to develop the native grains, and, since its presentation into Europe by Christopher Columbus and different pioneers, corn has spread to all spaces of the world suitable to its cultivation.

It is grown from 58° N latitude in Canada and Russia to 40° S latitude in South America, with a corn crop maturing somewhere in the world nearly every month of the year. It is the most important crop in the United States and is a staple food in many places.

The corn plant is a tall yearly grass with a bold, erect, strong stem. The huge narrow leaves have wavy edges and are dispersed alternatively on inverse sides of the stem. Staminate (male) blossoms are borne on the tassel ending the primary pivot of the stem.

Figure 1.1. History and origin of maize

Source: Image by Wikimedia commons

The pistillate (female) inflorescences, which develop to turn into the eatable ears, are spikes with a thickened pivot, bearing matched spikelets in longitudinal lines; each line of combined spikelets regularly delivers two columns of grain. Assortments of yellow and white corn are the most well known as food, however there are assortments with red, blue, pink, and black kernels, regularly banded, spotted, or striped.

Every ear is encased by adjusted leaves called shucks or husks. Numerous industrial varieties of corn are hereditarily altered for protection from the herbicide glyphosate or to create proteins from Bacillus thuringiensis (Bt.) to kill explicit insect pests. Also, a few strains have been hereditarily designed for more prominent drought resistance.

Business characterizations, in light of kernel texture, incorporate dent corn, flint corn, flour corn, sweet corn, and popcorn. Mark corn is portrayed by a downturn in the crown of the piece brought about by inconsistent drying of the hard and delicate starch making up the kernel.

> Flint corn, containing minimal delicate starch, has no downturn. Flour corn, made generally out of delicate starch, has delicate, coarse, effectively ground kernels. Sweet corn has wrinkled clear seeds; the plant sugar isn't changed over to starch as in other sorts.

Popcorn, a limit kind of flint corn portrayed by little hard bits, is without delicate starch, and heating makes the dampness in the cells extend, making the kernels to detonate. Upgrades in corn have come about because of hybridization, in view of crossbreeding of predominant ingrained strains.

Despite the fact that it is a significant food in numerous pieces of the world, corn is substandard compared to different cereals in dietary benefit. Its protein is of low quality, and it is inadequate in niacin. Diets in which it prevails regularly bring about pellagra (niacin-inadequacy infection).

Its gluten (elastic protein) is of nearly low quality, and it isn't utilized to deliver raised bread. It is generally utilized, in any case, in Latin American cooking to make masa, a sort of mixture utilized in such staple food varieties as tortillas and tamales.

Given that corn flour is without gluten, it can't be utilized alone to make rising breads. In the United States corn is bubbled or broiled on the cob, creamed, changed over into hominy (hulled portions) or feast, and cooked in corn puddings, mush, polenta, frying pan cakes, cornbread, and scrapple. It is additionally utilized for popcorn, desserts, and different produced cereal arrangements.

Corn is also used to produce ethanol (ethyl alcohol), a first-generation liquid biofuel. In the United States corn ethanol is typically blended with gasoline to produce "gasohol," an automotive fuel that is 10 percent ethanol.

In spite of the fact that corn-based biofuels were at first promoted as harmless to the ecosystem options in contrast to petrol, their production diverts arable land and feedstock from the human evolved way of life, starting a "food versus fuel" banter.

Cellulosic ethanol, which is produced using non - edible plant parts like rural waste, has a smaller impact on the food chain than corn ethanol, however the change innovation is by and large less proficient than that of original biofuels.

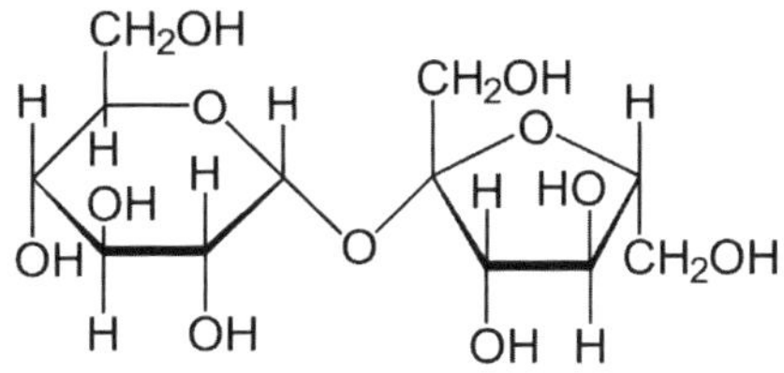

Figure 1.2. Structure of sucrose

Source: Image by Wikimedia commons

Numerous pieces of the corn plant are utilized in industry. Corn-starch can be separated into corn syrup, a typical sugar that is by and large more affordable than sucrose; high-fructose corn syrup is utilized broadly in handled food varieties like sodas and confections. Stalks are made into paper and wallboard; husks are utilized as filling material; cobs are utilized straightforwardly for fuel, to make charcoal, and in the planning of industrial solvents.

Corn grain is processed by wet milling, in which the grain is soaked in a dilute solution of sulphur acid; by dry milling, in which the corn is exposed to a water spray or steam; and by fermentation, in which starches are changed to sugars and yeast is employed to convert the sugars into alcohol. Corn husks likewise have a long history of utilization in the folk arts for items, for example, woven ornaments and corn-husk dolls.

1.1.1. Taxonomy

Maize is an individual from the grass family Poaceae (Gramineae), a classification it shares to numerous other significant rural yields, including wheat, rice, oats, sorghum, grain, and sugarcane. In view of fossil proof, it is assessed that these significant grass genealogies emerged from a typical predecessor inside the last 55 - 70 million years, close to the near end of the rule of dinosaurs. Maize is additionally coordinated in the genus Zea, a gathering of yearly and perennial grasses local to Mexico and Central America.

The genus Zea includes the wild taxa, known collectively as teosinte (Zea ssp.), and domesticated corn, or maize (Zea mays L. ssp. mays). For a long time, the connections inside genus Zea were the subject of much contention. The central difficulty in the taxonomy of maize and the identification of its closest relatives was the absence of a cob like pistillate inflorescence—or "ear"—in any other known plant.

The teosinte creates simply 6 to 12 kernels in two interleaved lines secured by a hard external Present day maize brags a cob comprising 20 lines or more, with various uncovered kernels.The teosinte is not similar to maize in the construction of its ear that nineteenth century botanists neglected to perceive the connection between these plants, putting teosinte in the variety Euchlaena as opposed to in Zea with maize (Doebley, 1990b).

Notwithstanding these significant actual contrasts, different morphological, cytological, and hereditary examinations at last outlined the connections inside variety Zea. H. G. Wilkes (1967) established the framework for the current classification plot in 1967 with the first careful monograph on teosinte.

Wilkes didn't endeavor a conventional pecking order yet rather introduced an arrangement of classification utilizing diverse geographic populaces, with independent racial assignments dependent on distinctive morphological highlights.

In 1980, Hugh Iltis and John Doebley (Doebley and Iltis, 1980; Iltis and Doebley, 1980) produced a system of classification that considered the probable evolutionary relationships between taxa. With the quantitative evaluation of numerous traits and the discovery of many additional populations, Jesus Sanchez (Sanchez G. et al., 1998) provided further characterization of this genus.

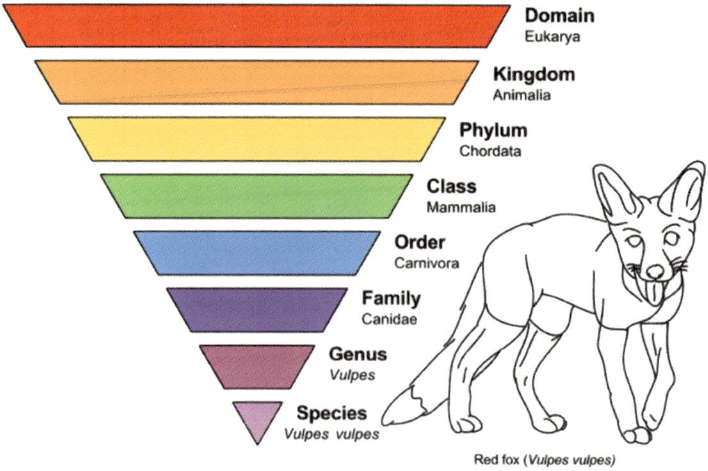

Figure 1.3. Taxonomic rank graph

Source: Image by Wikimedia commons

1.2. THE ORIGIN OF MAIZE

1.2.1. Historical Argument

As researchers toiled all through the mid-to late 1900s to put together an arrangement of classification for the family Zea, an equal riddle surfaced with respect to the beginning of maize. In spite of developing acknowledgment - reflected in the current scientific categorization - of the view that teosinte (Z. mays) is the quick progenitor of maize (Z. mays ssp. mays), agreement didn't come without any problem.

In the battle to comprehend the inference of the puzzling corn ear, two driving theories arose. In the last part of the 1930s, Paul Mangelsdorf and his partner Robert Reeves proposed a theory known as the three-sided speculation (Mangelsdorf, 1974; Mangelsdorf and Reeves, 1938, 1939).

This hypothesis expressed that maize was domesticated from some obscure wild maize, apparently a plant with structures that looked like the advanced maize ear. More specifically, as the name demonstrates, the theory comprised of three sections: A wild maize model from South America, which is presently either terminated or unseen, was the begetter of maize; teosinte is the posterity of a hybrid of maize and Tripsacum (another class of grasses); and segments of Tripsacum chromosomes had "sullied" maize germplasm.

In this way, Mangelsdorf and Reeves summoned a missing precursor to represent the limit morphological contrasts among maize and teosinte while depending on Tripsacum to clarify their similitudes. They highlighted their own fruitful cross of maize and Tripsacum as approval for their speculation.

Without a doubt, albeit the cross involved significant human mediation, Mangelsdorf and Reeves had the option to create a couple, generally sterile maize - Tripsacum half and halves. They likewise investigated backcross populaces of maize - teosinte mixtures and had the option to distinguish four variables (which they deciphered as four Tripsacum chromosomal sections) liable for the morphological contrasts among maize and teosinte.

For George Beadle, nonetheless, the morphological contrasts among maize and teosinte were not so enormous as to require a terminated predecessor. In June 1939, not exactly a year after the distribution of the three-sided speculation, he reacted with his own hypothesis on the beginning of maize, a thought he had persuaded himself regarding as a Cornell graduate understudy under the bearing of Rollins Emerson (Doebley, 2001).

In his teosinte speculation, Beadle (1939) expressed that maize is basically a trained type of teosinte. He accepted that through artificial choice by old populaces, a few little transformations with enormous impacts might have changed teosinte into maize.

Beadle really utilized Mangelsdorf and Reeves' own information against them, asserting that their four elements may very well to compare to four significant qualities, every one of which controlled a solitary characteristic that separated teosinte from maize.

He additionally tested their thought that a hybrid of maize and Tripsacum, which took such Herculean endeavors from them, would have happened in nature. By all accounts, these dueling speculations zeroed in on the beginnings of an unassuming ear of corn, yet at the center of the discussion was an issue more essential and maybe more extensive - a Darwinian discussion for the ages.

In one corner were developmental conservatives who held that advancement continues gradually after some time, through the amassing of numerous little changes in various qualities. Along these lines the sensational change from teosinte to maize was basically impractical in the simple 10,000 years in which people have been domesticating plants, and a more coherent beginning stage was required on which determination could act.

In the other corner were minds like Beadle's and Emerson's, the place where advancement could be faster whenever impelled by changes in a couple significant qualities. So, despite the fact that teosinte and maize may have looked strikingly changed, this distinction could be represented by just four or five significant qualities, clarifying why the two plants were generally hereditarily comparable (to such an extent that they could be effortlessly crossed to deliver rich posterity).

From its introduction in 1938 until the 1960s, the three-sided theory was generally acknowledged. Through useful coordinated efforts with noticeable archaeologists of his day (Mangelsdorf et al., 1964, 1967) and a side of the equator wide exertion focusing on maize germplasm preservation (Wellhausen et al., 1952), Mangelsdorf had the option to plug his hypothesis among a wide crowd, with his name getting inseparable from the investigation of maize development.

In the interim, Beadle briefly deserted his teosinte speculation for spearheading Nobel work on biochemical hereditary qualities and for the administration of the University of Chicago. During this time his restricting thoughts got little consideration.

Upon his retirement in 1968, in any case, Beadle re-joined the maize discussion, enthusiastically seeking after the question both on paper and face to face at a few gatherings specifically met to discuss the beginning of maize.

He came furnished with extra information that upheld his speculation (Beadle, 1972, 1977, 1980) and in the long run benefited from the waiting skepticism in the three-sided theory among numerous maize geneticists. Before Beadle's demise in 1989, a large group of scientific distributions had been given on the side of teosinte as the wild forebear of maize (Doebley, 1990a).

Figure 1.4. An illustration of maize – teosinte

Source: Image by Wikimedia commons

1.2.2. Modern Argument

The controversy continues. Although the mystery surrounding the origin of maize seemed to be solved, new pieces to the puzzle were added, given time and new technologies. Teosinte and its sister genus Tripsacum still take center stage in the modern argument, with one side steadfastly adhering to the teosinte hypothesis while the other revived the idea of a hybridization event.

1.2.3. Role of Huehuetenango in Evolution of Maize

The investigations completed toward the start of the century by Vavilov and others (Bukasov, 1981; Vavilov, 1997) showed that the hereditary variety of developed plants is moved in specific areas of the world, which they called "centers of beginning and variety".

Central America is one of the seven districts first recognized by Vavilov. Among the remarkable unique yields of this locale is maize, given the hereditary variety present in the area and the presence of wild species and subspecies identified with maize, an extraordinary on of which is teosinte.

The revelation of teosinte toward the finish of the most recent century pulled in the consideration of botanists in their endeavors to build up the starting points of maize. In 1939, Beadle exhibited dependably that maize and teosinte could be uninhibitedly crossed and that the half breeds acquired were totally prolific.

This recommended to him that the two taxa were conspecific and that they had as of late separated. Other scientists developed different opinions on the origin of maize. All these considerations omitted teosinte as a participant in the evolution of maize and proposed that it originated from a hypothetical "wild maize" (Mangelsdorf, 1974).

This researcher put together his hypothesis with respect to the way that the maize cob and the teosinte cob are massively extraordinary, which made it inconceivable over a brief period (around 10,000 years) for such developmental changes to have occurred. In any case, the greater part of the logical proof and chronicled references, identifying maize, support the theory that teosinte was a precursor of maize.

These contemplations are significant in the event that one understands that in Guatemala there are two species of teosinte right now classified as Zea and Zea luxurians appropriated in the Department of Huehutenango and the east of Guatemala individually.

This is the reason Guatemala is distinguished as one of the potential habitats of the beginning of maize in Central America. The first detailed studies of the diversity of maize in Guatemala were carried out in 1957, in which reference is made to factors that explain the evolution of maize, including the following:

Primitive maize included different varieties of palomero or small-grained maize whose differences were determined by different altitudes and different ecological and environmental conditions.

- The exotic maize landraces of South America established themselves on their own and became new hybrid cultivars by crossing with the old indigenous varieties.

- Through hybridization with teosinte, they came to form additional hybrid cultivars.
- Geographical and ecological isolation tended to preserve the landraces.
- The existence still today of a large indigenous population contributes to the conservation of the specific landraces of maize.

In Guatemala, 13 landraces and 12 neighborhood cultivars of maize have been recognized, of which seven landraces and four sub races are available in the Department of Huehuetenango, which shows the significance of the space for the hereditary variety of maize.

Effectively a few years prior (1945), McBryde had made an example variety of maize in Guatemala, the hereditary material of which was dissected and it was found, based on the hubs present in the chromosomes of material gathered in a space in the Department of Huehuetenango, that these contained "practically all the hub positions known in chromosomes of the multitude of races of maize on the planet".

They confirmed the presence of a large number of an alternate kind of maize in an exceptionally confined region. In this investigation, the investigators guarantee that western Guatemala is the place where most of the maize types filled in Central America, North America, and the marshes of South America, North America and the West Indies really originated.

The presence of Zea in the Department of Huehuetenango has come to be viewed as the main components in the generation of variability in maize. It was noticed that there was hereditary introgress among teosinte and the maize types filled in the adjoining area and the native populace perceived that hybridization happened.

In 1937, different mixture ears like teosinte were found; and then, in 1955, 45 F1 and 3 F2 maize-teosinte cross breeds were found nearby the Noyoya and San Antonio Huixtla towns. Doebly et al. (1987) made a comparative analysis of iso - enzymatic variations present in maize and teosinte using 13 enzymatic systems that codify 21 loci. This work included 56 teosinte populations represented in the geographic area of its distribution and 99 populations of maize distributed over Mexico and Guatemala.

Based on the outcomes, it was feasible to see that the Guatemalan teosinte disseminated over the Department of Huehuetenango (Zea huehuetenangensis) is the one that least takes after cultivated maize materials, though the teosinte of eastern Guatemala (Zea luxurians) bears a more noteworthy likeness to the MexicanZ. perennis and Z. diploperennis. Then again, it was expected that developed maize (Z. mays subsp. mays) was more identified with the teosintes of Mexican beginning, particularly with the taxon Z. Later examinations dependent on morphological perspectives, biochemical markers and atomic markers shows unmistakably that the hypothesis that maize started from teosinte is presently more by and large acknowledged. It is along these lines presently being recommended that maize is a tamed type of yearly Mexican teosinte Z.

However, if as some authors maintain, maize was introduced in Guatemala, evidence seems to indicate a very early introduction (at least some 4000 years ago). Time and the skill of the indigenous populations were sufficient to ensure the generation of the great morphological diversity present in this country.

1.3. THE NATURAL HISTORY OF MAIZE

Maize, likewise alluded to as corn or Indian corn in the United States and Great Britain, separately, is a cereal plant of the Gramineae group of grasses that today establishes the most generally circulated food plant on the planet.

Likewise, maize - from the Arawak mahiz - is filled in different districts and environments, from 58 degrees north scope in Canada and Russia to 40 degrees south scope in South America. Maize development and handling are driven by the production of food and domesticated animals feed, fermentation, and raw materials for industry.

Given its numerous utilizations, maize is probably going to be found in more than 1,000 items in a very much loaded U.S. store. The particulars of maize creation, generation, development, handling, and utilization - its flexibility, alterability, just as the recalcitrance of social and organic imperatives—keep on furnishing science with bits of knowledge into the past and conceivable eventual fate of the species. As anyone might expect, maize is the most examined plant species on earth.

1.3.1. The Ethnobotany of Maize

Ethnobotany, the investigation of advantageous connections between human societies and the plants on which they depend, is one of the numerous fields of study energizing examinations concerning the earliest domesticated maize and its resulting worldwide dissemination. Ethnobotanists, archaeologists, anthropologists, taxonomists, food and agricultural researchers, nutritionists, geneticists, biotechnicians, historians and numerous others are altogether attempting to discover answers to the various inquiries presented by the advancement and expansion of maize. This different collection of logical sources gives data about the characteristic and social history of maize.

Figure 1.5. An illustration of archaeologists of Hazara University

Source: Image by Wikimedia commons

1.3.2. The Quintessential Maize Plant

Specialists have set up that advanced maize developed from teosinte (God's corn), or Zea Mexicana, albeit a few botanists keep on arguing that it developed from an early Mesoamerican maize assortment called Chapalote.

Even the timing of maize origins has been questioned. A review of the botanical characteristics of both maize and teosinte will distinguish them and provide some idea of how maize developed into the fully domesticated cultigen it is today.

According to plant geneticist John Doebley, maize (Zea mays L. ssp. mays) and the teosintes (Zea spp.) differ profoundly in terms of vegetative characteristics and "inflorescence architecture" (1996, p. 66). That is, contrasts are explicit to the type of the plant and its reproductive design, including variation in the mode of development and course of action of the blossoms or sprouts along the axis of the plant. In this occurrence, the differentiations are generally prominent in the types of the male tassel, or spikelet, that sprouts at the summit of the maize or teosinte tail, just as in the structure and improvement of the female inflorescence or maize ear. Regardless of these, Doebley recognizes that maize and the Mexican teosintes are basically variations of similar organic species.

Therefore, they structured completely ripe half breeds and cross-fertilize; the intrinsic contrasts in chromosome structure and other hereditary angles among teosinte and maize are no more prominent than those saw among the different races of maize. Trials recognizing qualities of teosinte from maize have been reproduced as of late using molecular investigation.

The trademark reproductive example of maize is an extremely old and primitive one. Maize incorporates both male and female reproductive qualities and comprises an Andropogonoid grass that bears a spike-like axis or tassel on which blossoms or sprouts are connected and from which dust is scattered; it is a self-pollinating plant that scatters dust from the tassels to the "style" or "silk" of the maize ears (female inflorescences), where it is caught up in the reproductive cycle. While every tassel contains about 25 million dust grains, every female inflorescence or maize ear contains as much as 1,000 ovules or likely pieces.

Each pollinated silk is in this manner changed into an individual piece of maize that develops to contain a solitary ovule fundamental for the generation of the actual plant. Maize leaves track the daylight and ingest its energy; a field of maize is ideally intended for delivering significant returns from sunlight-based energy.

1.3.3. Maize Variation and Race

There are some 25 "essential" races of maize found in Mesoamerica, and none of these is unadulterated. The multiplication of mixture variations and ongoing advances in bioengineered or hereditarily altered varieties of maize has apparently fixed the destiny of this most old food item. Indeed, the worldwide multiplication of hereditarily designed food varieties is ready to totally uproot or supplant existing strains of the essential grain crops with naturally designed substitutes.

The developmental history and intrinsic alterability of maize are unpredictable to such an extent that researchers proceed to discussion and question the taxonomic identification of the multitude of surviving races of maize in both wild and homegrown settings. There is no understanding about the scientific classifications or quantities of races that may exist in any single world locale.

Since maize is so effectively hybridized, the quantity of assortments far surpasses some other yield species on record. Natural taxonomists have inexactly gathered these assortments into somewhere in the range of 300 races for the Western Hemisphere alone.

Early reading material scientific classifications distinguished just six races, including dent, flint, flour, sweet, pop, and waxy varieties. Of these, two dominate American commercial agriculture and incorporate the Flint (Zea indurata) and Dent (Zea indentata) varieties.

The nine significant sorts developed in the United States incorporate the Southwestern Sediments, Southwestern twelve-line, Pima-Papago, Great Plains Flints and Flours, Corn Belt Dents, South-eastern Flints, Southern Dents, Derived Southern Dents, and Northern Flints. Of those races of maize indigenous to Mesoamerica, four main groups of maize have been identified.

Their respective taxonomic classifications are based on the vegetative characteristics of the plant, characteristics of the spike or spikelet, characteristics of the cob, and the physiological, genetic, and cytological characteristics of the plants studied.

These primary maize groups are:

- Antique indigenous,
- Exotic pre-Columbian,
- Prehistorical mestiza, and the
- Not well-defined or modern races.

To this latter category may be added the proliferation of genetically engineered strains.

The antique native group comprises of those races that began in Mesoamerica with the primitive earliest races of maize. Varieties inside this gathering are believed to be proof of numerous autonomous starting points in different spaces of Mesoamerica. The races explicitly related to this gathering incorporate Palomero Toluqueño, Arrociclo Amarillo, Chapolote, and Nal-Tel.

1.4. MAIZE IN THE NEW WORLD

The historical backdrop of maize and its domestication might be followed back to around 8,000 years. Maize spread across the length and breadth of the Americas, and thusly to Europe, Africa, and Asia. Teosinte (Zea mexicana) has been connected with the earliest maize in Mesoamerica and was first reaped as early as 10,000 years ago.

The starting points of maize start on the Pacific incline of the cutting-edge Mexican provinces of Oaxaca, Tehuacán, and the Valley of Mexico. The earliest primitive corncobs found in Mesoamerica were acquired from specimens recuperated inside a cavern close to Oaxaca.

From that point maize diffused quickly into Central America and afterward into South America via the eastern inclines of the Andes roughly 4,000 years prior. Guatemala may have filled in as the source or channel for the selection of the earliest strains of maize in Andean South America and Peru.

Indeed, the underlying appearance of maize in Peru has been dated to 6070 B.C.E. Ecuador, Chile, and Argentina, with Andean Peru, structure an imaginable passage for the transmission of maize from Guatemalan sources into beach front valleys.

Nearly nineteen races of maize from ten Latin American nations have been related to the Classic time of 300 to 900 C.E. Six of these evidenced communications among Mesoamerican and Peruvian social orders from the most distant times of pre-Columbian social turn of events.

1.4.1. New World dispersals

Walton C. Galinat has followed the dissemination of maize into North America to a Northern Flint Pathway set up by 700 C.E. in the Rio Grande valley. From that point maize spread toward the north along both the eastern and western flanks of the Rocky Mountains, and toward the east along significant waterway courses - including those of the Arkansas, Mississippi, Platte, and Ohio Rivers - allowing it proceeded with development and dispersal toward the east.

By 1200 C.E., maize development was set up in upstate New York and New England. While the eight-paddled assortment of maize was developed in the southeastern United States in pre-Columbian occasions, the Southern Dent Pathway represents the appropriations of different assortments of maize after 1500 C.E., ensuing to Spanish contact in those districts.

Eventually, hybrid vigoro identified with the larger and more robust forms of maize can be followed back to the converging of the Northern Flint assortments with the Southern Dent assortments by U.S. ranchers of the Midwest during the nineteenth century.

As indicated by Galinat, this half breed combination can be described as having come about in an "accidental transformative explosion" that at last changed maize into a profoundly useful and significant food item.

1.5. MAIZE IN THE OLD WORLD

As per Sylvia Johnson, as early as the mid-sixteenth century, maize had been acquainted with Europe, western Asia, Africa, and China. These early Old-World experiences with maize were undecided toward the fascinating grain at that point known as Turkish wheat, Turkish grain, Spanish wheat, or Indian corn.

The negative responses to maize in the Old World to a great extent zeroed in on the conviction that maize was less nourishing than extant European grain items like the wheat, grain, or oat cereals utilized in the production of bread and related results.

Notwithstanding its impressive usefulness in correlation with wheat, its more limited developing season, and its significant versatile potential to minor conditions, maize was at first seen as a staple fit uniquely for creatures or the most unfortunate of the working class, who ground it up with water and ate it as a finely ground mush or porridge.

This came to be known as "polenta" to the people groups of northern Italy, which has since been consolidated into European and American foods. Starting in Italy, an assortment of fixings and added substances, including cheddar and pasta, have enhanced the elements of this helpless people's food and changed it into a worldwide top pick.

In China and Southeast Asia, maize is developed in rotation with other, more customary harvests like rice or millet, and sequential cropping (relay cropping) methodologies license a type of various editing that covers the existence patterns of at least two yields. These techniques made conceivable the age of harvest excesses in Asia past those initially related to the selective or conventional dependence on rice as an essential cultigen., The maize is being utilized all through Asia to enhance more customary harvests by broadening the developing season and growing production possibilities consistently. What's more, the production of maize feed and feed for animals has energized the selection of maize agribusiness all through the non-industrial nations of Asia and Africa. Maize gives the world's most cost-effective and highest yield plant asset presently accessible for the production of domesticated animals forage, grain, and feed.

1.5.1. Old World dispersals

The social, financial, and political effects of the European discovery of maize were apparent in the resulting populace boom that followed its presentation into the Old World. After 1492, maize quickly diffused into Europe, Africa, and Asia and was effective in enormous part since it didn't straightforwardly rival existing grain harvests like rice, wheat, oats, millet, and grain.

Maize was additionally fit to cultivation in otherwise helpless developing conditions identified with geology, soils, environments, aridity, and height. Essentially, maize additionally flourishes in outstandingly wet environments unsuited to wheat or generally bone-dry areas unsatisfactory to rice development. Besides, maize enjoys the extra benefit of quick returns and twice the useful yield per unit of place where there is wheat.

The selection of maize in Africa and China proclaimed a sensational social and social change. Maize gave a degree of food overflow that allowed the remarkable development of populaces. Though in Europe maize was viewed as an inadequate oat grain, fit distinctly for taking care of poor people and hungry and domesticated animals, in numerous spaces of Africa and Asia maize came to overwhelm the farming economies of numerous country states.

The usefulness and proficiency of maize agriculture and its low production and transportation costs made it a modest nourishment for slaves caught and held by European and Arabic slave brokers. Maize made conceivable the proficient and practical vehicle and trade of horrendous quantities of sub-Saharan Africans bound for the business sectors of Europe, the Middle East, and the Americas.

1.6. THE AFRICAN CONNECTION

In spite of the fact that it stays hazy who previously acquainted maize with Europe, Africa, and the Old World, various researchers currently contend that the Portuguese provinces of Africa filled in as the initial conduit to the dissemination of maize around there.

Jean Andrews asserts that maize, beans, peppers, squash, and turkeys diffused into the Balkans, or southeastern Europe, via Portuguese Africa, India, and the Ottoman Empire in the period following the journeys of Columbus (1993, pp. 194–204).

So significant was the effect of maize on the African economy that, similar to Mesoamerica, culture and society, means and settlement, political economy and gender relations, and the particular cooking styles and culinary advancements of every one of these immense locales were quickly changed to oblige the selection of maize and those human diasporas with which it was related.

The novel maize-based social complex of agrarian practices, broad settlement examples, and capacity, conveyance, and food preparing advancements related to maize development in fact filled a significant part of the change being referred to.

In Africa's Emerging Maize Revolution, Derek Byerlee and Carl Eicher acknowledge that the adoption of maize has been the primary engine driving the transformation of the African social, political, and economic landscape for the many societies that have been swept up in this new agricultural revolution.

Besides, in numerous spaces of Africa, a large part of the customary African agrarian complex, which are fixated on such harvests as millet, has been uprooted by maize. This has a lot to do with the changing substance of African cooking at the most major degree of examination, and all the more for the most part, at the interface of social change and change.

Africans prepare a maize porridge - called kpekple in Ghana, bidia in Zaire, sadza in Zimbabwe, putu in Zululand, mealie in South Africa, and posho or ugali in East Africa - consumed by millions. Virtually no African country has remained untouched by the diffusion and exchange of maize, and the agricultural practices on the African continent range from the simple sowing of maize kernels along rivers and streams to the cultivation of maize in household gardens.

While far and wide, these customary practices are primitive contrasted with the extent and power of agribusiness advancement and interest in economically suitable maize farming field frameworks.

1.6.1. The Maize Harvest of the Machine Age

One need just travels to places like the province of Nebraska to understand that maize farming dominates the rural customs of certain social orders. Indeed, a drive through Nebraska during the developing season may leave a few outsiders with the effect that it comprises of a consistent, apparently perpetual, and exceptionally thick field of maize. For as long as 200 years, farmers and farming researchers in such regions have fostered an assortment of means, innovations, and crossovers appropriate for the proceeding with proliferation of maize.

In The Story of Corn (1992), Fussell summarizes the many agricultural technologies, cropping and harvesting methods, hybrids, commercial products, and cultural and religious values identified with maize agriculture in the Americas and other parts of the world.

Strategies for reaping of maize range from the hand culling of hybrid maize cobs from home nurseries to transcending innovative consolidates or collecting machines and farm haulers manipulated—in the biggest joins - to separate maize ears or cobs at the pace of twelve columns all at once and a huge number of bushels each day.

Indeed, even a more seasoned combine or harvester can gather nearly 10,000 bushels of maize each day, yielding 150 to 200 bushels for every section of land. Then again, the support and upkeep of such machines effectively runs into the great many dollars each year, a sum liable to twofold or significantly increase in the Third World.

A new combine harvester can cost from $100,000 to $200,000 or more in the United States. Whereas subsistence farmers throughout much of the Third World continue to thresh maize by hand without specialized equipment or resources, this task is left to agribusiness giants and commercial agricultural concerns in industrialized nations.

An array of farm machinery patents for the harvesting and processing of maize appeared shortly after the industrialization of farms. The U.S. Patent Office Report of 1860 lists hundreds of patents for corn planters, cultivators, harvesters, cornhuskers, corn shellers, cornstalk cutters, corn-shock binders, cornstalk shocking machines, corn cleaners, seed drills, rotary harrows, smut machines, corn and cob crushers and mills, and seed drills (Fussell, p. 144).

The approach of the canning business in 1862 and the multiplication of new land-grant universities dedicated to science, farming, and industry helped fuel the industrialization and modernization of maize obtainment, preparing, capacity, circulation, and hybridization. in time, the industrialized cultivating tasks and horticultural cooperatives of the Corn Belt received many motorized strategies for collecting and sifting maize.

Before long the towering grain silos and high-rise grain elevators of the Midwest supplanted the unassuming corncribs received from the North American Indians. In spite of this, customary family corncribs endure the invasion of the Industrial Age, and stabilities incorporate the Mesoamerican cuezcomatl (thatch - roofed adobe brick granary), the bunk logged storage facilities of the Sierra Tarahumara Indians of northern Mexico, and the mud lined maize grain storehouses of Africa.

In each case, resource farmers and agribusiness giants alike should consider the troubles of putting away maize at ideal conditions and balance humidity, the moisture content of the kernels, and the potential for pest infestations.

1.6.2. Maize: The Columbian Exchange

Rather than discovering gold in the New World, Columbus was one of the main Europeans to see maize (corn) (Morison, 1991). Maize was just found in the New World until Columbus brought it into the Old World. Columbus carried maize to the Old World and fundamentally changed the existences of Europeans (Nunn and Qian, 2010).

Maize was appealing to the Old - World inhabitants for several reasons (Nunn & Qian, 2010). It helped to improve diets by providing much needed nutrition and calories (Nunn & Qian, 2010). Crosby (1972) writes, "Few other plants produce so much carbohydrate, sugar, and fat" (p. 171).

Maize was a food source from the New World that brought a change to the diets of Europeans. They now had access to and were able to eat a food that had significant nutritional value.Maize was easily grown in the Old World (Nunn & Qian, 2010). Crosby (1972) says, "Maize will produce good crops in an extreme variety of climates" (p. 171).

Maize had the option to grow in regions that were excessively wet for wheat and excessively dry for rice. The harvest was additionally turned with wheat, so Europeans could now grow two yields in a year rather than one harvest (Crosby, 1972).

Maize develops rapidly and produces an enormous yield. Along these lines, maize is as yet depended on by Europeans as a significant food source.

In China, maize was found to fill in regions not appropriate for different yields. In the eighteenth and nineteenth centuries, China had a populace blast that constrained individual to live in slopes and mountains.

These individuals found that maize could develop where land was once infertile (Crosby, 1972). In this time when individuals were moving to live in the slopes and mountains, floods continued happening in the valleys of China washing away supplements for rice fields (Mann, 2011).

Rice farmers attempted to grow a beneficial harvest because of these floods. Chinese authorities understood the floods were the reason for disintegration since the common scene was changed to plant and develop maize.

Authorities restricted the development of maize in the mountains and slopes, yet the boycott was not authorized. Despite the fact that rice farmers were losing farm benefits, they were gathering tax-exempt lease from individuals living in the mountains and slopes.

Since the rice farmers would prefer to gather tax-exempt lease than have a beneficial harvest that one should pay burdens on, the restriction on developing maize was rarely implemented (Mann, 2011). The boycott ceased to exist and these days, China is one of the top makers of maize (Crosby, 1972).

However, depending on maize as a significant food source hit a tough situation. A cold spell hit Europe during the 1550s, which made Europeans ease off from depending on it so much. These days, a great many Europeans eat a diet dependent on maize.

Romania and Yugoslavia are the world's biggest maize makers on the planet (Crosby, 1972). In southern Africa, maize has become the essential food hotspot for workers (McNeil, 2008).

Relying on maize for human food is almost world - wide and relying on maize as animal food is just as vast. In the United States, most cows, pigs, chickens, and turkeys all eat maize as part of their diet (Crosby, 1972). Buckets and buckets of chopped and ground maize are used to feed barns full of animals every day. Therefore, maize has not only become relied on by humans for food but is also relied on as food for animals.

With so much dependence these days on maize, one considers what life would resemble in the Old World without it. Europeans would not have a particularly nutritious food source, so maybe the populace would be smaller because of less sustenance.

The livestock in the Old World might be smaller and consequently offer less meat because of less sustenance in their weight control plans. History would be unique if Columbus had not carried maize to the Old World. Maize changed the Old World. It changed the scene in China and individuals had a more nourishing nourishment for themselves and their creatures.

1.7. THE FUTURE OF MAIZE

Maize is prepared into a bewildering cluster of consumer products going from natural corn and popcorn to cornstarch, corn oils, car fills, like ethanol and gasohol, and cocktails, including corn beer (chichi) and bourbon. In addition to the around 1,000 maize-based items that one is probably going to be discovered in the neighborhood grocery store, the hereditarily altered results of maize are making their own culinary variety and possibilities, pitfalls, controversies, and complications for the universe of food production and biotech enterprises.

Past the variety inherent in the production and circulation of maize in the commercial center, actually maize stays the main horticultural harvest for more than 70 million farm families around the world. A lot of the world's farmers who develop maize are in agricultural countries of the Third World.

This reality that the combined with the "hereditary disintegration" of the yield, has incited some to find out if maize can be reared to guarantee the maintainable advancement of the harvest (Sevilla, p. 221). On the other hand the claims for patent encroachments against farmers by agribusiness organizations and biotechnology firms for the unlicensed utilization of their licensed half and halves are any sign, the possibilities of hereditary variety and crossover life once related to maize might be compelled to ever less and progressively more vulnerable hybrid offspring.

1.8. CONCLUSION

Maize is one of the world's most significant yields – answerable for around six percent of human calorie consumption. Past human utilization it has a variety of employments, including for use as animals feed, for maturation for the American liquor industry, and for production of plastics and ethanol. Around the world, around a billion tons of maize is created each year.

The history of maize is controversial, with scientists, historians and archaeologists proposing competing origin theories. One thing that is widely accepted is where maize comes from.

Despite the fact that maize has for quite some time been a critical harvest on the European landmass, it just got huge in the UK moderately as of late, where it is primarily developed as nourishment for animals, and as game cover.

Moreover, in late many years the expanding pattern for green power has driven farmers to utilize it as an energy crop for anaerobic digesters. The measure of maize filled in the UK has soared since the 1970s. In 1973 there was around 20,000 sections of land of maize being cultivated in Britain, by 2016 this had arrived at 450,000 sections of land. Also, there is no indication of this development easing back down.

REFERENCES

1. Drury, L., 2019. THE SURPRISING HISTORY OF MAIZE - Bright Maize. [online] Bright Maize. Available at: <https://www.brightmaize.com/surprsing-history-maize/> [Accessed 28 June 2021].
2. Edward S. Buckler IV and Natalie M. Stevens, 2005. Maize Origins, Domestication, and Selection. [ebook] Available at: <http://www.bionica.info/biblioteca/Buckler2005MaizeOrigins.pdf> [Accessed 28 June 2021].
3. Encyclopedia, 2019. The Natural History of Maize | Encyclopedia.com. [online] Encyclopedia.com. Available at: <https://www.encyclopedia.com/food/encyclopedias-almanacs-transcripts-and-maps/natural-history-maize> [Accessed 28 June 2021].
4. FAO, n.d. The role of women in the conservation of the genetic resources of maize. [online] Fao.org. Available at: <http://www.fao.org/3/y3841e/y3841e04.htm> [Accessed 28 June 2021].
5. The Columbian Exchange, n.d. Maize. [online] The Columbian Exchange. Available at: <https://thecolumbianexchange.weebly.com/maize.html> [Accessed 28 June 2021].
6. The Editors of Encyclopaedia, 2021. Corn. [online] Britannica. Available at: <https://www.britannica.com/plant/corn-plan> [Accessed 28 June 2021].
7. Wagley, 1957
8. Valladares, 1957
9. López-Austin, 1994
10. Mangelsdorf Cameron, 1942
11. Kempton and Popenoe, 1937
12. Randolph, 1955
13. Doebley, 1987
14. Weissinger et al., 1983
15. McClung de Tapia, 1992
16. McNeish, 1992
17. Doebley, 1990
18. Doebley and Iltis, 1980
19. Iltis and Doebley, 1980

20. Mangelsdorf, 1974
21. Mangelsdorf and Reeves, 1939
22. Doebley, 2001
23. Beadle, 1939
24. Mangelsdorf et al., 1964, 1967
25. Wellhausen et al., 1952
26. Beadle, 1980
27. Bukasov, 1981
28. Vavilov, 1997
29. Mangelsdorf, 1974
30. McBryde, 1945
31. Doebly et al., 1987
32. Columbus, 1993
33. Fussell, 1992
34. Morison, 1991
35. Nunn & Qian, 2010
36. Crosby, 1972
37. Mann, 2011
38. Crosby, 1972
39. McNeil, 2008

Chapter 2

Land Preparation Practices for Maize Production

CONTENTS

2.1. Introduction 28

2.2. Methods Involving No Tillage or Seedbed Shaping 29

2.3. Animal versus Tractor Power: Some Considerations for the Small Farmer 35

2.4. Animal-Drawn Disk Harrow 38

2.5. Beans Grown on Raised Beds 39

2.6. Gear for Seedbed Shaping 41

2.7. Step-by-Step Guide to Planting Maize 43

2.8. Land Preparation Methods for Maize Production 46

2.9. Conventional Till Flat Planting 48

2.10. Impacts of Land Preparation 49

2.11. Conclusion 51

References 52

This chapter deals with the land preparation practices for the production of maize. It describes the various methods involved in it like the methods involving tillage which incorporates the primary and secondary tillage. It also mentions the size of the seeds, the depth of the tillage.

It goes on explaining the handling of crop residues, flat seedbeds and raised seedbeds. The mid of the chapter explains the winkle planting, gear for seedbed shaping. It also talks about the physical land preparation and mechanical field preparation.

It tells different types of land preparation methods for maize production and land readiness strategies. It defines the raised bed or edge planting, zero-till planting. Towards the end it also illustrates the conventional till flat planting and furrow planting.

2.1. INTRODUCTION

While setting up another date plantation, certain activities should be executed to guarantee long term achievement of the plantation. One of these activities include initial land preparation which ought to be done before relocating of the plant material (branches or tissue culture-determined plants).

The reason for land preparation is to give the vital soil conditions which will upgrade the fruitful foundation of the youthful branches or the tissue culture plants got from the nursery. Considering the idea of the date palm, one cannot "save" on this activity and expectation for long haul manageability of the farm.

The aim is to enable the date grower to plan and structure the implementation process in advance, ensuring the successful establishment of the date plantation. Planning forms part of the initial preparation and will help to limiting unnecessary stoppages during the implementation phase.

Land is a comprehensive idea (FAO, 1976). It is a delineable space of the world's earthbound surface, enveloping all ascribes of the biosphere quickly above or underneath the surface, including those of the close surface environment, the dirt and landscape frames, the surface hydrology (counting shallow lakes, rivers, marshes and swamps), the near surface sedimentary layers and related groundwater hold, the plant and animal populaces, the human settlement design and actual consequences of at various times human movement, for example, terracing, water stockpiling or waste constructions, streets, structures, and so on (FAO, 1995).

Soil capacities are considered to be the living space, contribution for creation, environment controller, stockpiling, waste and contamination regulator, hydrologic and biotic natural capacity, and as file and legacy (FAO, 1996). Every one of these capacities requires an appropriate nature of land to be satisfied.

Land characteristics are perplexing properties that influence the appropriateness of the land for a predetermined use in an unmistakable manner (FAO, 1976). Land quality is the state of land comparative with the necessities of land use, including horticultural creation, ranger service, protection and natural administration (Pieri et al., 1995).

Land quality pointers (LQI) are utilized progressively to give advantageous depictions of present status or state of the asset and to measure its exhibition, and foresee reactions. Indicators are insights or measures that identify with a condition, change of value, or change in state (FAO, 1976).

In this context, three types of indicators are distinguished: pressure indicators, state indicators and response indicators (FAO, 1996). The LQI programme has been developed to harmonize the combined objectives of food production and environmental protection (FAO, 1996).

They are proposed to address the significant land-related issues of public and worldwide importance, for example, land use pressures, land corruption, and soil and water preservation, as well as strategy related inquiries. It was intended to be utilized for strategy and program detailing at different levels to advance and screen innovations, strategies and projects to guarantee better utilization of normal assets and economical land the executives.

Land Preparation for Cropping

On small farms, land preparation methods for the reference crops may or may not involve actual tillage (working the soil with hoes, plows or other equipment) or seedbed shaping (leveling land or making raised beds or ridges).

2.2. METHODS INVOLVING NO TILLAGE OR SEEDBED SHAPING

Under states of shifting cultivation, low administration or steeply slanting or rough soils, land is frequently cleared by basically slashing and/or burning, trailed by making the seed openings with a planting stick or hoe. No endeavor is made to really work the dirt or to shape a particular kind of seedbed.

- **Slash, burn and plant:** This strategy is generally reasonable for sandy soils which are normally free or for different soils that are kept up in good filth (a free, brittle condition) by a long vegetative decrepit which produces soil humus. It very well might be the lone doable technique for rough soils or those with articulated inclines where culturing would speed up disintegration.
- **Slash, mulch and plant:** This strategy is fit to similar conditions. The vegetation is cut down or executed with an herbicide and afterward left on a superficial level to shape a mulch (a defensive covering). The seeds might be planted in the ground or may even be dissipated over the ground prior to slashing.

The mulch is significant for disintegration and weed control, rationing soil dampness, and keeping soil temperatures more uniform. The International Institute for Tropical Agriculture (IITA) has discovered this framework extremely useful for maize and cowpeas and has created two kinds of hand operated planters equipped for sowing seed through a mulch.

There isn't anything fundamentally inappropriate with both of these strategies. At times, culturing and seedbed molding may enjoy some significant benefits:

- Soils prone to drainage problems due to topography, soil conditions or high rainfall usually require the use of raised beds or ridges for successful crop production (except for rice).
- If liming is needed to correct excessive soil acidity, it must be mixed thoroughly into the top 15-20 cm of soil to be fully effective.
- Chemical fertilizers containing phosphorous and potassium and organic fertilizers should be incorporated several centimeters into the soil for maximum effectiveness. Under non-tillage methods, they can still be correctly applied using a hoe or machete, but it is definitely more work.
- Chemical fertilizers containing phosphorus are best applied to the reference crops in a band 7.5-10 cm deep that parallels the crop 5-6 cm to one side. A fertilizer furrow can be made easily with a wooden plow or other animal drawn implement.
- Most animal- or tractor-drawn planters require a tilled seedbed for successful operation. There are exceptions, however, such as the IITA planters.

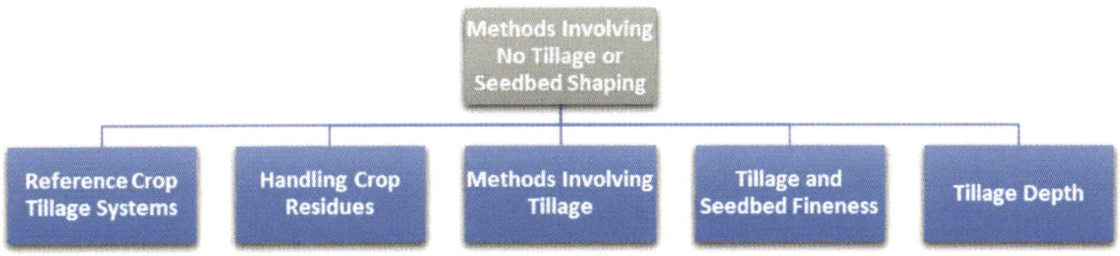

Figure 2.1. Methods involving no tillage or seedbed shaping

2.2.1. Methods Involving Tillage

Tillage refers to the use of animal- or tractor-drawn equipment or hand tools to work the soil in preparation for planting and has five main purposes:

- To break up and loosen the soil to favor seed germination, seedling emergence, and root growth
- To chop up and/or bury the previous crop's residues so they will not interfere with the new crop
- To control weeds (an ideal seedbed is completely weed free at planting time)
- To incorporate (mix into the soil) liming materials and fertilizers (chemical or organic)
- To shape the kind of seedbed most suited to the particular soil, climate, and crop (i.e., raised beds, ridges, flat seedbeds).

Primary tillage alludes to the initial breaking up of the dirt by plowing or utilizing a heavy-duty digging hoe. Depth of plowing for the most part goes from around 15-30 cm, contingent upon the sort of plow utilized, its traction source, and the soil.

For instance, a bull drawn wooden plow won't have the infiltration capacity of a tractor drawn moldboard plow, particularly in hefty soils. Secondary tillage alludes to any extra tillage activities among plowing and planting to separate clods, cut up weeds, and smooth out the seedbed. It is most generally performed with some kind of harrow (a carry out used to pummel and smooth the dirt).

Secondary tillage is shallower than planting and requires less force. Ridging and bedding (framing edges or beds for raised planting) additionally can be included for this category.

2.2.2. Reference Crop Tillage Systems

The reference crops share similar essential tillage methods; however, these fluctuate with the specific soil, the accessible tillage equipment, and the requirement for joining lime or compost. There are three fundamental tillage frameworks, each with benefits and disservices:

- **Plow (or hoe)/Plant:** If plowed at the right dampness level, a few soils (particularly topsoil and sands) might be reasonable for planting without any secondary tillage to break up the clods. Most soils can be hand-planted in the wake of plowing, since the farmer has preferable power over seed depth over when a mechanical planter is utilized.

  He can likewise shove any huge clods to the aside or split them up while walking down the row. This sort of unpleasant seedbed is in reality invaluable in terms of weed control since the cloddy surface debilitate their development. It additionally favors dampness penetration and lessens overflow. Then again, if bedding or ridging is required, a superior occupation should be possible if any huge clods are first separated by harrowing (cultivating).
- **Plow/Harrow/Plant:** This is the most common system where animal- or tractor-drawn planters are used, unless the soil breaks up well enough under plowing alone. If soil conditions are conducive to weed growth, the ground should be harrowed as close to planting as possible to give the crop a head start on the weeds.
- **Minimum Tillage:** Farmers with admittance to work vehicle or animal drawn tillage equipment may exaggerate tillage, particularly through rehashed harrowing to control growing weeds or separate clods. Killing one harvest of weeds by stirring the soil just invigorates another by drawing other weed seeds nearer to the soil surface.

Unnecessary culturing invigorates the microbial breakdown of humus and may additionally annihilate great soil physical condition by over-pulverizing the soil. The apparatus, animal, and people walking through additionally compact the soil, hindering root development and seepage.

> Tillage is extreme when hand devices are utilized to plan ground for the reference crops, in light of the measure of work it would include. Cut and-consume and slice and-mulch techniques fall under zero tilling, as do strategies utilizing uniquely adjusted mechanical planter to plant seed into unplowed ground (normal in the U.S.).

2.2.3. Tillage and Seedbed Fineness

The degree to which clods need to be broken up depends mainly on seed type and seed size and whether hand planting or mechanical planting will be used.

- Seed type: Maize, millet and sorghum are monocots with seedlings that break through the soil with a spike-like tip. This reduces the need for a clod-free seedbed. Peanuts and other pulses are dicots, and emerge in a blunt form, dragging the two seed leaves with them; they tend to have more trouble with clods.
- Seed size: Large seeds have more strength than little seeds, empowering youthful shoots to push all the more viably through harsh seedbeds. Maize seeds are huge monocots, which gives them particularly great clod handling capacity.
- Peanuts and other pulses are huge seeded, yet this benefit is partly offset, since they are dicots. The little seeds of sorghum and particularly millet is less powerful, yet this is counterbalanced by the way that they are monocots. Little seeds require shallower planting than bigger ones, and cloddy soils don't permit this sort of exactness if mechanical planter are utilized.
- Farmers can generally get by with cloddier seedbeds when hand planting. They have more power over planting depth and can shove any huge clods to the aside. Also, it is not unexpected under hand planting to plant a few seeds for every opening, which allows them a superior opportunity of breathing through.

Clayey soils, particularly those low in humus, are normally in a cloddier condition subsequent to furrowing than loamy or sandy ones. Most furrowing happens toward the finish of the dry season, when soils are extremely dry, which emphasizes the issue. Precipitation following furrowing may fundamentally decrease lump issues on certain soils by separating the clods.

2.2.4. Tillage Depth

A plowing depth in the 15-20 cm range is usually adequate, and there is seldom any advantage in going deeper. In fact, shallower plowing is often recommended for low rainfall areas like the Sahel to conserve moisture.

In some areas, tractor-drawn sub-soilers (long slender shanks that penetrate right down to 60 cm) are used in an attempt to interrupt up deep hardpans (compacted layers). Effects are honest to terrible, relying at the form of hardpan; the ones which includes a dense clay layer regularly re-cement themselves inside a short time.

Approximately 60 – 65 percent of the reference plants' roots are discovered within the topsoil, due to the fact that this sediment is more fertile (partially due to its better organic be counted content) and less compacted than the subsoil.

However, any roots that enter the subsoil can make use of its valuable moisture reserves, creating a crucial distinction in the course of a drought. Right fertilization of the topsoil will encourage lots deeper root improvement.

On the other hand, bad drainage and excessive acidity within the subsoil will avoid or prevent root penetration.

2.2.5. Handling Crop Residues

There are three basic ways of handling the previous crop's residues (stalks, leaves, branches) when preparing land: burning, burying and mulching:

- Burning - This destroys the organic matter contained in the residues, but may be the only feasible solution where suitable equipment is lacking or where time is short.
- Burying - chopping residues up with a disk harrow or slasher and then plowing them under is a common practice in mechanized farming.
- Mulching - Chopping up residues and leaving them on top of the ground has some definite benefits such as greatly reducing soil erosion caused by rainfall and wind as well as water losses due to evaporation. However, there are two disadvantages to mulching which should be considered:
- Residues are left on the surface and can interfere with the operation of equipment such as planters, plows, and cultivators which may plug up.

- Mulching is not recommended for peanuts, especially in wet regions, since they are very susceptible to Southern stem rot (Sclerotium rolfsii), which can incubate on unburied residues from any type of plant.

2.3. ANIMAL VERSUS TRACTOR POWER: SOME CONSIDERATIONS FOR THE SMALL FARMER

Within the developing countries, tractor energy and its associated device are especially constrained to massive farms and to regions where labor charges are excessive. The large investment, gas and repair costs, and preservation necessities all weigh closely towards the purchase of such equipment by way of small farmers.

Spare components and the vital restore facilities are usually lacking, which means that a breakdown can be disastrous. A study by ICRISAT at the economics of tractors in India showed new evidence that they significantly boom yields, cropping intensity, timeliness or gross returns consistent with hectare. Cash can generally be tons better spent on animal traction equipment, advanced seeds, fertilizers, and other high return inputs.

However, there are situations wherein tractor strength may be justified:

- Animal-drawn gadget won't be enough to fulfill the production wishes of the intermediate farmer who has approximately five-20 ha of land. In this situation, small horsepower device can be very appropriate. The global Institute for Tropical Agriculture (IITA) farming structures software has developed a five hp gasoline powered multipurpose gadget unit that may plant subject vegetation with a -row "punch" planter, haul 500 kg in a trailer, and convert to a walk-in the back of tractor for rotary tillage, ridging, brush slashing, and plowing rice paddies. Other styles of low-horsepower gadgets are available from different producers.
- The small farmer can sometimes benefit by hiring tractor work on an as-needed basis during peak periods when his normal labor supply is insufficient to meet demands.

2.3.1. Basic Tillage Equipment for Plowing and Harrowing

Hand Implements: Heavy duty burrowing cultivators can be powerful for little regions. In Kenya, for instance, essentially all little possessions are arranged thusly, albeit a normal family can't deal with significantly more than 0.5 ha with this technique.

In a wet-dry environment, most land readiness happens when the soil is hard and dry, which stances added obstacles for hand devices. Some extension services suggest that land be set up toward the finish of the previous wet season before the soil dries out. Be that as it may, this isn't generally conceivable because of standing harvests.

2.3.2. A Digging Hoe and a Heavy - Duty Hoe Blade

One normal kind of wooden plow. The greater part of them has metal tips to decrease wear.

Wooden Plow

Designs of wooden plow goes back to numerous hundreds of years. They regularly are animal drawn, and some have a metal tip. They don't modify the soil or cover crop deposits yet fundamentally make grooves through the soil. Their adequacy relies a ton upon soil type and dampness content. The depressions they make additionally can fill in as seed and manure furrows.

Moldboard Plo$w

This is the best plow for turning under grass, green compost harvests, and hefty yield buildups, for example, slashed up maize stalks. It likewise covers weed seeds further and harms enduring weeds more than other gear. Moldboard plow are accessible in animal drawn models (normally only one plow base) and work vehicle models (generally two to six bottoms). Contingent upon plow size (width of the moldboard as seen from the front or back) and soil condition, they will enter to 15-22 cm.

Except if outfitted with a spring trip gadget, moldboards don't deal with rough soils well. They are not too fit to drier regions as plate plows. They likewise experience issues in tacky earth soils and may frame a plow skillet (a slim compacted layer that can obstruct root development) whenever utilized at a similar depth a seemingly endless amount of time after year.

Disk Plow

Better fit than the moldboard to hard, clayey, rough or tacky ground, yet doesn't cover deposits as viably. This is a benefit in drier regions where surface buildups lessen wind and water disintegration and cut down dampness evaporation. Disk plows are not suggested for nut ground where Southern stem decay is an issue, since surface plant deposits harbor the spores.

They additionally won't work really hard turning under grass turf. Disk plows are primarily accessible in tractor drawn models. Dissimilar to moldboards plows, they are more averse to shape a plow pan whenever utilized at a similar profundity year after year.

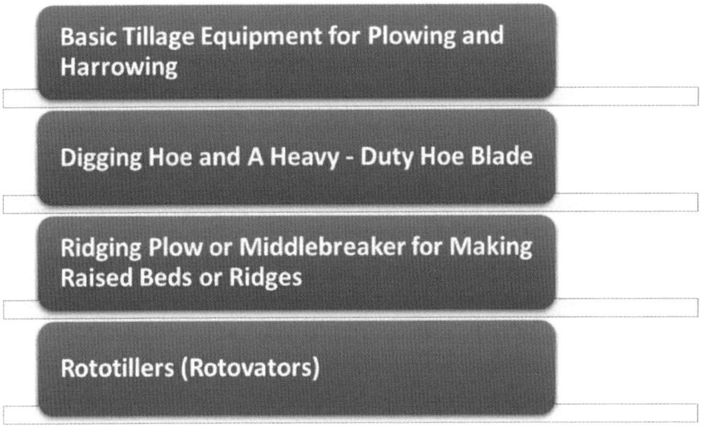

Figure 2.2. Animal versus tractor power: some considerations for the small farmer

2.3.3. A Ridging Plow or Middle breaker for Making Raised Beds or Ridges

Ridging Plows (Lister Plows or "middle breakers") comprise of a twofold sided moldboard that tosses soil the two different ways. This will deliver a progression of rotating furrows (trenches) and ridges when worked over a field. Contingent upon the environment and soil, the harvest is either planted in the furrows (in low precipitation regions with no seepage issues) or on top of the edges (in high precipitation regions or those with waste issues). Such wrinkle planting is beneficial in drier regions for cereal harvests, since it preserves dampness. Soil is tossed into the column as the season advances for weed control, and this additionally sets the roots profound into the dirt, where dampness is more satisfactory. Such wrinkle planting isn't suggested for peanuts and frequently not for beans because of expanded root decay and stem decay issues.

2.3.4. Rototillers (Rotovators)

These are accessible in tractor powered models. They thoroughly beat the soil and halfway cover crop deposits. Substantial models can be utilized for a quick overview complete culturing position. The detriments are that power prerequisites are extremely high and the soil can be effectively exhausted with this carry out.

Indeed, rototillers do an undeniably more exhaustive occupation of seedbed planning than is required for the reference crops and are best utilized for vegetable ground. A rototiller or rotavator. Note the rotating edges in the engine behind the wheels.

2.4. ANIMAL-DRAWN DISK HARROW

2.4.1. Disk Harrows

Disk harrows are usually utilized in the wake of plowing to separate lumps, control weeds, and smooth the dirt prior to planting. They are likewise used to slash up coarse harvest buildups prior to furrowing (particularly if a moldboard or circle plow will be utilized), yet heavier models with scalloped (plates with huge serrations) are best for this reason.

Both animal and farm hauler drawn models are accessible however they are costly and inclined to visit bearing disappointment except if routinely lubed. Large, heavy duty versions pulled by tractors are often called Rome plows and can sometimes substitute for plowing.

The gangs of disks are offset to the direction of travel so that they cut, throw, and loosen the top 7.5-15 cm of soil but pack down the soil immediately below that. Repeatedly harrowing a field prior to planting can actually leave it harder than before plowing if done when the soil is moist.

2.4.2. Spike-Tooth Harrows

These comprise of a metal or wood frame studded with stakes or spikes; additional load as stones or logs might be required under certain conditions for more adequacy. They are utilized to smooth the seedbed and separate lumps (at the right dampness content), and are particularly appropriate for killing little weed seedlings that may arise prior to planting.

Spike-tooth harrows are made in numerous widths and are grouped by weight and the length of the prongs. At times, this sort of harrow can be run over the genuine harvest lines from a few days subsequent to planting up until the seedlings are a couple of centimeters tall to control early geminating weeds or to separate any dirt crusting. Spike-tooth harrows will stop up if refuse is left on the dirt surface.

There are two models of a spike-tooth harrow, as mentioned below:

- **Spring-Tooth Harrows:** These have tines produced using spring steel that burrow, lift, and slacken the top 7.5-10.0 cm of soil, separate blocks, and smooth out the seedbed. Both creature and farm hauler drawn models are accessible. They are not fit to hard or horrendous ground however handle stones well.
- **Field Cultivators:** These are comparable in appearance to chisel plows, however ordinarily are not as built as weighty. They can be utilized for starting culturing on ground with minimal surface buildup, however are fundamentally utilized as an optional culturing execute for weed control. Most models are intended for tractor use.

2.5. BEANS GROWN ON RAISED BEDS

Seedbed Shape

The best seedbed shape depends more on the climate and soil involved than on the particular reference crop.

2.5.1. Flat Seedbeds

This shape is utilized where soil dampness is sufficient for crop development and where there are no waste issues. Under such conditions, the reference crops are frequently planted on a flat seedbed and afterward "hilled up" with (soil is moved into the harvest column and mounded around the plants) as the season advances to control in-line weeds, offer help, and improve seepage. In warm conditions, muggy regions where stem decay is an issue, this training isn't suggested for peanuts.

2.5.2. Raised Seedbeds (Ridge or Bed Planting)

Under heavy rainfall and additionally helpless seepage, the reference crops are typically planted on ridges or raised beds to hold them back from getting "wet feet". This same limits soil-borne infection issues like root spoils and assists control with watering disintegration if the ridges are run on the shape.

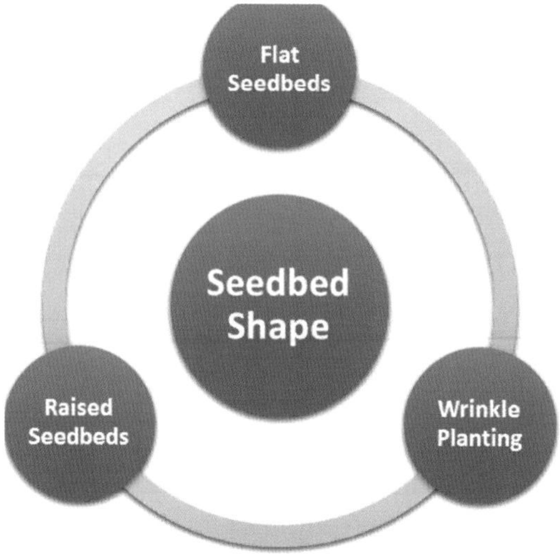

Figure 2.3. Various shapes of seedbed

Water invasion is empowered and spill over limited. Moreover, ridge planting makes for simpler section of burrowing gear when peanuts are collected. At long last, more soil is accommodated crop development under this framework.

The fundamental hindrance of ridge planting is the speed up loss of soil dampness from the hills – and typically it is not a major issue in wet regions besides during droughts. In drier regions mulching would be gainful. In areas where the wet season begins moderate, the harvests might be flat planted and afterward "hilled up" as the downpours increment. Wrinkle water system consistently requires ridge planting.

2.5.3. Wrinkle Planting

Under states of low precipitation or helpless soil water-holding limit (i.e., sandy soils), crops are frequently planted in the wrinkle base between ridges where soil dampness is more noteworthy. Soil would then be able to be tossed into the furrows to control in-line weeds and improve seepage (if precipitation gets) as harvest development advances. This sort of depressed planting isn't suggested for peanuts in sodden regions, since it empowers stem and root decays, especially if soil is tossed into the column.

Note: Local farmers normally have great seedbed experience, so be careful with altering dependable techniques without first thinking about every one of the points and running a few preliminaries.

2.6. GEAR FOR SEEDBED SHAPING

Flat seedbeds typically require no unique endeavors past plowing and potentially nerve racking. The extra land evening out is required.

Ridges or beds can be made with burrowing diggers, extraordinary ridging furrows (see culturing gear segment) or farm vehicle drawn circle bedders (moving plates organized at restricting points to hurl soil to shape beds). The yield can be planted either on top of the ridges or in the furrows, contingent upon the soil and environment.

2.6.1. Physical Land Preparation

When a reasonable region for building up the estate is chosen and the arranging activity is finalized, the genuine readiness can be enacted. These exercises are partitioned to construction and speed the execution cycle to be prepared for planting at the most appropriate time, as per the specific local climatic conditions.

2.6.2. Mechanical Field Preparation

The mechanical or initial soil preparation concerns mainly the preparation of a fi eld for further detailed preparation such as irrigation system installation, hole preparation, etc. Actions, if applicable to the area, include:

- Debushing/bush clearing;
- Removal of stones and rocks;
- Ripping; and
- Levelling of the soil.

2.6.3. Irrigation System Installation

The kind of water system framework to be utilized will be controlled by the accessibility of water, geographical and soil conditions. At the point when the underlying soil planning is finished, the establishment of the necessary water system framework will be carried out as per the endorsed configuration.

Soil Improvement

The scheduling of the soil improvement program will rely upon the date cultivator, as specific applications could be joined with the underlying activities of soil arrangement. Because of the long holding up period, planting to first creation, it's anything but a pattern to set up date ranches on new soils, except for regions where date palm is utilized for intercropping.

If new soils are considered, the soil improvement programme will mostly deal with:

- The application of organic matter; and/or
- The elimination of soil salinity.

2.6.4. Organic Material

In general, most soils are poor in organic matter content and the improvement of this situation plays an important role in soil fertility. Some of the advantages of a higher humus content in the soil are summarized as follows:

- Enhances crumb formation which improves the respiration of the roots;
- Increases the water infiltration rate;
- Increases the water holding capacity;
- Lowers soil compaction and crust formation; and
- Limits the harmful effects of alkalinity and improves the leaching of salts.

2.6.5. Salinity

In an attempt to reclaim salt affected soil, consideration should be given to:

- The type of salinity/alkalinity,
- The drainage possibilities of the soil profile,
- The origin or the source of salts,
- The quality of irrigation water and
- The leaching of salts from the soil.

If the source of salts is identified as drainage water from higher lying areas, a cut-off canal may be sufficient to eliminate this source of "salt" supply.

Poor seepage typically goes together with soil saltiness issues and subsequently the improvement of the waste potential ought to be tended to before any draining system is carried out. A soil cover (mulching) and the use of natural material will improve the water penetration bringing about improved seepage (barring soils with obstructive layers). In saline soils (dissolvable salts present as chlorides, sulfates and additionally carbonates of calcium, sodium or magnesium), just filtering will be important to deplete the overabundance salts. On account of antacid as well as saline-soluble soils, sodium can be supplanted through the utilization of gypsum or acidifying specialists like sulfur. When the sodium has been supplanted, a program ought to be followed to filter it out.

At the point when the water system water is of low quality, appropriate seepage and over water system, without the advancement of a water table, is vital.

2.7. STEP-BY-STEP GUIDE TO PLANTING MAIZE

Maize is a costly harvest to develop, costing around £300/section of land, so hitting the nail on the head and accomplishing great yields are significant.

Maize Growers Agronomist Simon Draper and consultant Ian Grandfield, of Green Gem Agriculture, provide step-by-step guidance on growing maize this spring including:

- Field selection
- Seed bed preparation
- Drilling
- Variety selection
- Weed control
- Nutrition
- And dealing with pests and disease.

2.7.1. Step 1: Field Selection

There are three key aspects of field selection:

- **Altitude:** Maize is a sub-tropical plant and the altitude at which it is grown is critical to its ability to mature.

  Wherever the crop is grown above 140m there must be other factors to favor of growing at this height, such south-facing and free-draining soil.
- **Aspect:** Aim for south-facing sites that are not too exposed.
- **Soil type:** Maize needs a dry soil in the spring, but not a soil type that will dry out too much in late spring when the young plants are developing.

Developing maize on substantial, earth type soils is presumably the greatest impairment to attempting to get the best out of this yield due to the trouble in making the right seed-bed in the spring.

Medium soils that are not difficult to work in the spring, free-depleting and will heat up right off the bat in the season to empower the youthful plants to get the most ideal beginning are the most preferred.

Exceptionally free-depleting and light, sandy soils can be excessively outrageous.

2.7.2. Step 2: Seed-Bed Preparation

The aim is to achieve a deep, loose seed-bed. Maize is a free-rooting plant so there needs to be no restriction on root development in the early stages.

Crops that struggle with compaction then struggle to mature and achieve optimum yield.

Farms with heavy soils often need to winter plough to achieve the correct seed-bed in the spring. But if there is a wet March, water can hold under the furrows and can take a long time to dry out.

The soil at the top can look dry and great, yet it's essential to look further down and see what conditions are truly similar to.

Crumblers and other equipment that will develop the ground to accomplish the receptiveness of the soil that maize needs for planting ought to be utilized instead of any pack that will minimize the bed. Having the option to kick a foot through soils is a decent aide.

2.7.3. Step 3: Drilling and Sowing

Maize seed should be drilled to a uniform depth and into moisture.

Soil temperatures should by 8C first thing in the morning for four to five consecutive days for maize to grow.

The last 10 days of April should see all maize drilled, but climate will obviously depend on location.

There is an old adage that says for every day that the crop is sown into the month of May there will be a 1% loss in yield.

With seed medicines prone to be restricted one year from now (2020), farmers should penetrate further to keep away from rook harm. Seed should be more profound than a rook's bill thus ought to be planted into the seedbed in any event 7-8cm.

To guarantee right planting profundity, cultivators should check behind the bore and reveal at any rate one-meter length of seed to guarantee penetrating profundity and separating are uniform.

At the point when seed is drilled further the seedbed is probably going to be colder so producers ought to anticipate deferred germination. To counterbalance this, it very well might merit taking soil temperatures at 7-8cm profound and boring later into hotter soils.

Maize seed can come through from a more profound planting, yet in the event that the dirt conditions imply that it's anything but quite a while to arise, and afterward comes through into harsh spring climate, there will be misfortunes.

A wide range of drills have been assessed for maize throughout the long term, yet most producers return to exactness drills and 76cm columns. Maize needs penetrating gradually with one seed conveyed each 11-13cm.

2.7.4. Step 4: Choosing a Variety

This should be based on experience of what has already been grown or what is being grown successfully on farms where conditions are similar.

Yield is paramount for dairy farmers growing forage maize and must be a high priority in variety choice.

A modest target is 49.9t/ha (20t/acre) fresh weight. Always select a variety that has the potential to give the level of maturity needed by the end of October.

2.7.5. Step 5: Weed Control

Once drilled, all fields should receive a pre-emergent herbicide to combat weeds.

This will also mean post-emergence herbicides have less work to do.

Don't delay weed control, once they reach the 2-leaf stage they will start out-competing maize.

Maize is very poor at competing with weeds in its earliest stages of development so this must be avoided.

2.7.6. Step 6: Crop Nutrition

Nitrogen is the driver and a lot of crops just don't get enough because of an over-estimation of what the slurry can provide.

Maize requires:

- 180kg/ha of potash
- 40kg/ha of phosphate
- And 150kg/ha of Nitrogen

Although these will vary depending on soil indices or where large quantities of FYM or slurry have been applied there should be no requirement for P and K.

Therefore, farmers need to know precisely what nutrients have been applied by slurry or FYM.

Ideally, the application of nitrogen should be split with 75% in the seedbed and the remaining 25% applied at the 1-3 leaf stage. This will help the crop get through its yellow phase often seen at emergence.

2.7.7. Step 7: Pests and Diseases

When maize follows grass there can be pest problems such as wire worms, but growers need to be aware that leather jacket and frit-fly sprays are banned after 31 March.

Although maize is a relatively pest-free crop to grow, the removal of these treatments may have an impact in the future.

European corn-borer has been a problem in the South West and eye-spot can also occur in some crops.

Insecticide dressings can be applied at drilling and there is a fungicide dressing to prevent rotting of the newly drilled seed.

These dressings break down after 21 days so there can be a risk where there is slow emergence of seed in cold, wet seed-beds.

2.8. LAND PREPARATION METHODS FOR MAIZE PRODUCTION

Land readiness strategies extraordinarily impact development and yield boundaries of maize and soil properties. The decision of a technique relies upon the vegetation cover and the way in which the soil surface is to be uncovered for planting of seeds is subject to the thickness of weeds. The essential land arrangement strategies for maize production are regular (furrow and harrow) Tillage and protection or no culturing. Khurshid et al., (2006) detailed that among the yield production factors, Tillage contributes up to 20%. Protection Tillage includes the utilization of cleaver, cultivator, pickaxe, herbicide application or mulch culturing. Mulch Tillage leaves crop buildup on the dirt surface for fast germination and agreeable yield. The utilization of regular Tillage tasks is hurtful to soil, henceforth, there is a critical interest and accentuation on the shift to preservation Tillage and no-Tillage strategies to control soil disintegration (Iqbal et al., 2005).

Land planning on business ranches is finished by farm vehicle drawn carries out where early furrowing preceding the beginning of the downpour is trailed by a couple of frightening, yet this training is changing because of the significant expense of working apparatus and the trouble in getting extra parts to try different things with diminished and zero Tillage (Raemaekers, 2001).

2.8.1. Tillage and Crop Establishment for Maize

Tillage and harvest foundation is the key for accomplishing the ideal plant stand that is the principle driver of the harvest yield. In spite of the fact that the harvest foundation is a progression of occasions (cultivating, germination, development and last foundation) that relies upon cooperation of seed, seedling profundity, soil dampness, strategy for planting, hardware and so on at the same time, the technique for planting assumes an essential part for better foundation of yield under a bunch of developing circumstance.

Maize is principally planted straightforwardly through seed by utilizing various strategies for Tillage and foundation yet during winters where fields are not stay empty on schedule (till November), relocating should be possible effectively by raising the nursery. Nonetheless, the planting technique (foundation) fundamentally relies upon a few components viz the perplexing communication over the long haul of cultivating, soil, environment, biotic, apparatus and the executives season, trimming framework, and so forth As of late, asset protection advancements (RCTs) that incorporate a few practices viz. zero culturing, least culturing, surface cultivating and so forth had come by and by in different maize based trimming framework and these are practical and climate cordial. Subsequently it is vital that various circumstances require distinctive planting strategies for accomplishing better return as depicted underneath:

2.8.2. Raised Bed (Edge) Planting

By and large the raised bed planting is considered as best planting technique for maize during rainstorm and winter seasons both under overabundance dampness just as restricted water accessibility/rainfed conditions. Planting/ planting ought to be done on the southern side of the east-west edges/beds, which helps in great germination.

Planting ought to be done at appropriate dividing. Ideally, the raised bed grower having slanted plate, measuring or roller type seed metering frameworks ought to be utilized for planting that works with in situation of seed and composts at legitimate spot in one activity that aides in getting great harvest stand, higher usefulness and asset use proficiency.

Utilizing raised bed planting innovation, 20-30 % water system water can be saved with higher usefulness. Besides, under impermanent abundance soil dampness/water logging because of weighty downpours, the furrows will go about as seepage channels and harvest can be saved from overabundance soil dampness stress.

For understanding the maximum capacity of the bed planting innovation, perpetual beds are fitting wherein planting should be possible in a solitary pass with no preliminary culturing. Perpetual beds are more useful under overabundance soil dampness circumstances as the penetration rate is a lot higher and yield can be saved from the impermanent water logging injury.

2.8.3. Zero-Till Planting

Maize can be effectively developed with no essential Tillage under no-till circumstance with less expense of development, higher ranch productivity and better asset use effectiveness. Under such condition one ought to guarantee great soil dampness at planting and seed and manures ought to be put in band utilizing zero-till seed-cum-compost grower with wrinkle opener according to the dirt surface and field conditions.

The innovation is set up with enormous number of ranchers especially under rice-maize and maize-wheat frameworks in peninsular and eastern India. Nonetheless, utilization of fitting grower having appropriate wrinkle opener and seed metering framework is the key of accomplishment of the no-till innovation.

2.9. CONVENTIONAL TILL FLAT PLANTING

Under weighty weed pervasion where synthetic/herbicidal weed the board is uneconomical in no-till and furthermore for rainfed regions where endurance of harvest relies upon moderated soil dampness, in such circumstances level planting should be possible utilizing seed-cum-compost grower.

2.9.1. Furrow Planting

To prevent evaporative losses of water during spring season from the soil under flat as well as raised bed planting is higher and hence crop suffers due to moisture stress. Under such situation/condition, it is always advisable to grow maize in furrows for proper growth, seed setting and higher productivity.

2.9.2. Transplanting

Under serious cropping systems where it's anything but conceivable to empty the field on schedule for planting of winter maize, the odds of deferred planting exists and because of postpone planting crop foundation is an issue because of low temperature so under such conditions relocating is another option and grounded procedure for winter maize.

Along these lines, for the circumstance where fields are abandoned during December-January, it is prudent to develop nursery and relocate the seedlings in wrinkles and apply water system for ideal harvest foundation. Utilization of this strategy helps in upkeep of worldly separation in corn seed production regions for production of unadulterated and great quality seed just as quality protein maize grain.

For planting of one hectare, 700 m2 nursery region is required and the nursery ought to be raised during second fortnight of November. The time of seedlings for relocating ought to be 30-40 days old (contingent upon the development) and relocate in the period of December-January in wrinkles to get higher usefulness.

2.10. IMPACTS OF LAND PREPARATION

2.10.1. Impacts of Land Preparation on Soil Physical Properties

Customary culturing impacts soil porosity, mass thickness, infiltration opposition and dampness content (Khurshid et al., 2006). Yearly unsettling influence and pounding of soil through furrowing and frightening produce a better and free - soil structure contrasted with preservation or no-culturing strategies which leave soil unblemished (Rashidi et al., 2008).

However, the use of tractor tends to compact the soil to increase penetration resistance and bulk density (Unger and Kaspar, 1994). Penetration resistance over 1000kPa typically diminishes yield (Khalilian et al., 1991) and causes the start of root development reduction (Ishaq et al., 2001). Soil compaction may fundamentally weaken usefulness of soil by diminishing air circulation, soil water stockpiling and yield water use productivity while more noteworthy soil entrance, mass thickness and lesser absolute porosity is found on no culturing than plowed soils during maize development (Cassel et al., 1995; Griffith et al., 1996).

The zone of maximum soil bulk density roughly corresponds to the depth of tillage (Pikul and Asae, 1995) while the optimum soil bulk density for growth and yield of maize ranges from 1.40 - 1.68 Mgcm-3 (Czyz, 2004). Conservation tillage impacts soil actual properties, for example, mass thickness, invasion and water maintenance.

No culturing diminishes water misfortune from soil and improves soil dampness systems than furrow and harrow (Azooz et al., 1996) on the grounds that more surface spaces of soil are presented to daylight and stream of wind in traditional culturing.

No-culturing keeps up surface buildups, limits soil unsettling influence, empowers develop of natural material, protects the dirt design, and preserves soil water. Land arrangement strategy that consolidates natural matter into the dirt expands air circulation through tunneling and deterioration exercises of soil living beings (MacRobert et al., 2007).

2.10.2. Impacts of Land Preparation on Maize Performance

The customary and protection tillage techniques fundamentally impact yield and yield segments of harvests with ordinary tillage strategy recording essentially better return contrasted with no tillage (Rashidi et al., 2010).

Maize root entrance is more noteworthy in expectedly plowed soil than under no-work condition (Nitant and Singh, 1995). Roots are accumulated more at the surface layer under no-work conditions than routinely plowed soil however are more plentiful at more profound layer in customarily plowed soils than no-tillage, (Mauryaa and Lal, 2008).

Moreover, continuous ploughing results in plough pan which restricts nutrient movement and root penetration (Unger and Kaspar, 1994; Iqbal, 2006). Gul et al. (2009) reported the effect of tillage practices on biological yield of maize to be significant with conventional tillage producing higher biological yield than no-tillage.

There was a critical impact of tillage on leaf region per plant and leaf region list of maize. Leaf region per plant and leaf region file was less under no culturing when contrasted with ordinary tillage (Karunatilake and Schindelbeck, 2000).

2.10.3. Impacts of Land Preparation on Weeds

As indicated by Lampkin (1998), ploughing will in general cover weed seeds beneath the profundity from which they are fit for developing. Tillage alone cum great editing strategies is regularly the awesome most monetary technique for weed control (Lal, 1979).

Tillage straightforwardly influences the seed bank by actually blending the soil (Ball and Miller, 1990) and helps in overseeing herbicide opposition weeds that contend with developing harvests to decrease yield (Anderson, 2004 referred to by Chokor et al., 2008).

Nonetheless, changes from regular to preservation culturing practices can cause shifts in weed species and densities (Wilson and Foy, 1990 referred to by Jones et al., 1999). Gul et al. (2009) showed that new weed biomass is higher in the zero - culturing contrasted with customary culturing.

As horticultural land is ploughed and harrowed, weed biomass and time needed for weeding are decreased (Elliot et al., 1993). Moreover, lush weeds are more under zero culturing contrasted with regular culturing plots. Hand weeding twice all through the production season delivers least weed biomass since hand weeding successfully controls weeds (Syawal, 1998; Khan et al., 1998).

2.11. CONCLUSION

This chapter gives insight to the reader about the land preparation practices for the production of maize. It puts emphasis on different types of method such as the methods involving no tillage or seed back shaping, methods involving tillage. It also highlights the reference crop tillage systems.

It then defines the tillage and seed back fineness together with the tillage depth along with handling crop residues. It also explains the considerations for the small farmers under the topic animal versus tractor power. It throws light on the significance of basic tillage equipment for ploughing and harrowing. It illustrates the factors upon which beans grows on raised beds and how the best seedbed shape depends more on the climate and soil involved than on the particular reference crops. It sheds light on the gears for seedbed shaping, physical land preparation, mechanical field preparation, irrigation system installation. It talks about the different land preparation methods for maize production including the raised bed planting method, zero-till planting, conventional till flat planting and in the end, it describes the furrow planting.

REFERENCES

1. Agridaksh.iasri.res.in. n.d. Tillage and crop establishment. [online] Available at: <http://agridaksh.iasri.res.in/html_file/maize/Tillage%20and%20crop%20establishment.htm#:~:text=Maize%20is%20mainly%20sown%20directly,successfully%20by%20raising%20the%20nursery.> [Accessed 28 June 2021].
2. Hunt, J., 2016. Step-by-step guide to planting maize - Farmers Weekly. [online] Farmers Weekly. Available at: <https://www.fwi.co.uk/livestock/step-step-guide-planting-maize> [Accessed 28 June 2021].
3. Nzdl.org. n.d. Traditional field crops: Planning and preparation: Land preparation for cropping. [online] Available at: <http://www.nzdl.org/cgi-bin/library.cgi?e=d-00000-00---off-0hdl--00-0----0-10-0---0---0direct-10---4-------0-0l--11-en-50---20-about---00-0-1-00-0-0-11-1-0utfZz-8-00&a=d&c=hdl&cl=CL1.17&d=HASH7d1d4cf32e41fd9a480587.8.2> [Accessed 28 June 2021].
4. P. Klein and A. Zaid Date, n.d. Chapter Vi: Land Preparation, Planting Operation And Fertilisation Requirements. [online] Fao.org. Available at: <http://www.fao.org/3/Y4360E/y4360e0a.htm> [Accessed 28 June 2021].
5. Samuel Appah, 2012. Land Preparation Methods And Weeding Frequency Effects On Soil Properties And Maize Performance. [ebook] Department of Agricultural Engineering. Available at: <http://129.122.16.11/bitstream/123456789/5886/1/MSc.%20THESIS%20SAMUEL%20APPAH.pdf> [Accessed 28 June 2021].
6. Teklu Erkossa Jijo, 2005. Land Preparation Methods and Soil Quality of a Vertisol Area in the Central Highlands of Ethiopia. [ebook] Institut Für Bodenkunde Und Standortslehre. Available at: <https://d-nb.info/976411008/34> [Accessed 28 June 2021].

Chapter 3

Climate and Soil Requirement of Maize Farming

CONTENTS

3.1. Introduction 54

3.2. Agroclimatology of the Crop 56

3.3. Climatic Requirements 58

3.4. Weather Stress in Maize Cultivation 61

3.5. Soil: The Foundation of Agriculture 64

3.6. Soil Degradation and Crop Production 69

3.7. Sustainable Soil Management 70

3.8. Soils for Maize 71

3.9. Soil Requirements for Growing Corn 72

3.10. Conclusion 75

References 77

This chapter gives insights to the reader about the climate and soil requirement in context of maize farming. It begins with illustrating the agroclimatology of the crop, the climate requirements which includes factors like the temperature, rainfall and humidity, sunshine and the wind.

It throws light on the various aspects of climate such as the precipitation and the requirements of water. It also talks about the weather stress in maize cultivation, soil fertility and crop growth as well as the soils for agriculture, soil organic matter. It also puts emphasis how farming adjusts the regular cycling of supplements in soil.

It focuses on soil requirements for growing corn such as the corn soil needs, corn climate needs, and corn soil nutrients. In the end, it highlights how the effects on climate like the global warming, supply of water in the future generation will influence the farming of maize.

3.1. INTRODUCTION

Environment inconstancy impacts all areas of the economy, yet farming stays the most defenseless. Usefulness of yields shows changes starting with one year then onto the next, being fundamentally impacted by the inconstancy of climatic conditions and, specifically, the production of outrageous climatic occasions.

Accordingly, in Romania, on about 64% of the arable land, the soils are influenced, to a more noteworthy or lesser degree, by incessant dry spells, over long and surprisingly back-to-back years, separately by an overabundance of dampness in stormy years in excess of 6 million ha).

The effect of climatic changeability on the development, advancement and arrangement of farming yields is evaluated by the capability of meteorological boundaries to guarantee ideal vegetation conditions or produce ominous impacts relying upon the level of force of the upsetting element, the mode and term of activity, the likelihood of production and return in existence, just as the weakness of plant species to the production of outrageous occasions.

Each physical, synthetic and organic interaction that decides the development and advancement of horticultural yields is directed by explicit climatic necessities, and any deviation from these prerequisites verifiably decides the fluctuation of agrarian produce. Corn is quite possibly the main harvests on our planet. Corn has ventured into crops because of specific plant and natural highlights (Bîlteanu Gh., Bîrnaure V.).

In addition to other things, corn is well open minded to dry spell and warmth, is described by couple of illnesses and few nuisances, can be developed on various reliefs and under various climatic conditions, bears the monoculture, it's anything but a pruning crop, leaving the weeds clean, is a magnificent forerunner to wheat, utilizes natural and mineral composts, makes it simple to acquire exceptionally useful half and halves and adjusts to development regions, responds emphatically to water system, and so forth

Maize is a plant that has a decent versatility to extreme lighting conditions and high daytime temperatures with low photo responsiveness. The fundamental climatic factor of corn crop is temperature. The warm necessities of the plant are moderately high all through the vegetation period, in spite of the fact that they are very unique in relation to one vegetation stage to another.

The ideal maize development temperature is 28-30ºC, yet it likewise utilizes moderate temperatures. It is accepted that in the development, blossoming and filling periods of the grain, temperatures of 10°C reason plant development to stop, yellowing or brightening of the plant.

High temperatures above 32°C are considered as basic temperatures for corn. Through the perspiration cycle, the plant loses an extremely huge measure of water, drying out, regardless of whether the plant has a supply of water in the soil.

At higher temperatures (48°C for 6 hours) and under relatively low humidity conditions (below 30%) the maize plant is irreversibly dry. As it advances in vegetation, the requirements of maize are rising as against temperature. To ensure a good growth rate, average temperatures should not fall below 13ºC in May and below 18ºC in June, July and August.

At the point when the temperature is high during the day and low during the evening, the development pace of maize plants is diminished, the developing season is delayed and the production diminishes. Of the country's horticultural region, on 76% of the space the warmth factor is positive and entirely ideal for maize crops, which can guarantee exceptional returns.

The amount of heat, however, varies greatly in agricultural areas, very distant averages of annual average temperatures from less than 7ºC to more than 11ºC give a high degree of temperature variation in the period of growth and development of maize plants in different areas. The main climatic factors that determine the level of vegetal production, including corn production, are: heat (temperature), light and water (precipitation).

Of the climatic factors involved in plant growth and the formation of agricultural production, light and natural heat are inseparable elements, the values of which generally follow the same curve. Therefore, temperature and precipitation were used in calculations and determinations.

3.2. AGROCLIMATOLOGY OF THE CROP

Environment is interrelated with other production factors and ought to be seen either as an asset to be overseen or a factor that should be controlled. Reasonable utilization of soil, capital and work ought to be offset with utilization of environment and climate data. The reaction of the maize harvest to environment relies upon the physiological cosmetics of the mixture/assortment being developed.

Yield contrasts are the consequence of the hereditary arrangement of the half breed, the ecological conditions under which the harvest is developed, and the invasion by crop bothers (infections, insects and weeds). The last yield will rely upon: mixture choice; soil fruitfulness; soil water; and control of harvest pests.

These components will be examined in a specific order. Every farmer needs to choose the mixtures that are generally reasonable for the climatic locale wherein his homestead is found. In calm districts of the world the length of the ice - free season directs the crossovers that are reasonable, on the grounds that the maize plant can't withstand temperatures underneath about – 2°C.

Developing degree-day and warmth unit ordering frameworks have been created for most mild districts of the world, so the developing time needed by every maize cross breed can be coordinated with developing degree-day or warmth unit evaluations for the ice-free developing season in each climatic locale.

In tropical areas of the world, it is the stormy season beginning that directs the determination and ideal chance to plant maize half and halves. They should be chosen to coordinate with the anthesis time frame to when the soil dampness is probably going to be generally satisfactory, except if there is water accessible to flood the yield when fundamental.

The dirt richness needs to have an ideal equilibrium of the three significant supplements, nitrogen, phosphorus and potassium, and the essential micronutrients, like boron, calcium, magnesium, manganese and molybdenum. The farmer ought to have the dirt on his homestead evaluated for the three significant supplements each little while to decide how much compost ought to add to boost maize creation.

This is a significant administration practice in both calm and tropical areas. The farmer ought to know about treatment necessities and strategies concerning the soil supplement levels, the development stage, the harvest assortment, the designated yield and the agronomic practices. To amplify maize yields soil dampness ought to be kept up above half of the accessible water limit in the establishing profundity of the dirt profile all through the developing season.

This isn't generally conceivable in either mild or tropical climatic locales as precipitation can be exceptionally scant and irregular in certain years. Notwithstanding, it is fundamental to in any event have satisfactory soil dampness at the hour of anthesis to have a full arrangement of bits on the ear at collect time.

This is when supplemental water through water system would be generally useful. Harvest contamination and pervasion by sicknesses, insects and weeds can altogether decrease yields in both calm and tropical locales and agrometeorology is significant in crop insurance.

Sicknesses are best controlled through maize reproducing programs that foster cross breeds with protection from such infections as: leaf curses (of which Bipilaris, Colletotrichum and Excesohilum strains are normal models); root and tail decays (of which Phytophera, Fusarium, Gibberella and Diplodia strains are agent cases); rusts (of which Puccinia and Polyspora are plain delineations); and mucks, (for example, Ustilag and Spacelothian strains). Maize sicknesses are generally not constrained by showering with fungicide synthetic compounds, then again, actually seed is regularly treated prior to planting with a fungicide powder to control soil microbes that harm the incipient organism before germination. (These parasites incorporate Pythium, Fusarium, Diplodia, Rhizoctonia, Penicillium species). Wet and cold soil (<10°C) conditions at planting time is generally positive for these seed illnesses.

> The problem of disease and pest control among different production levels is particularly acute in the small-scale, resource-poor systems under which maize is typically grown in tropical regions of the world. The most inexpensive control measure for insects is through crop rotation, which ensures that maize is not grown on the same land year after year.

Various types of weeds can overrun maize harvests and cause yield misfortunes in both mild and tropical areas. The Amaramthus, Panicum, and Butilon species are the most hindering in mild areas and the Striga species in tropical locales, particularly in Africa. Among the 23 Striga species in Africa, Striga hermonthica is the most unfavorable.

In mild districts more often than not weeds are constrained by herbicide application. Since these synthetics are applied not long after planting maize, when the dirt should be soddened however not very wet, climate conditions assume a significant part in the achievement of weed control in mild environments.

Some weed species have created safe strains to the 'triazine' synthetics that are normally applied before rise of maize seedlings, so it is at times important to apply elective herbicides after rise. Herbicides are regularly excessively costly for farmers to use in tropical districts, so other social practices, for example, crop turn and hand development, are utilized to control weed invasions.

Climate conditions assume a part as well, yet are not as basic with respect to herbicide application. According to several studies around the world (James et al. 2000; OMAFRA, 2002; Dogan, 2003) the best time to minimize the effect of weeds on maize yields is within 4 to 8 weeks after planting when maize is in the 2 to 8 leaf stage. When weeds are controlled culturally during this initial period of maize growth, then shading by the crop is quite effective in controlling weed growth during the remaining time to maturity.

3.3. CLIMATIC REQUIREMENTS

Vegetables began in various pieces of the world: some came from tropical or subtropical nations and others from calm zones; some from damp regions and others from more bone-dry environments. Every sort of vegetable has its own ideal development necessities, with some more particular, and others less so.

Reproducing and determination of new cultivars have considered a more noteworthy flexibility to less good developing conditions than was conceivable previously, however the intrinsic climatic prerequisites of a particular sort of vegetable have not changed tangibly.

Figure 3.1. Climatic requirements for maize production

3.3.1. Temperature

Temperature is the most important climatic factor to be considered in vegetable production. It determines when and where a certain crop can be grown, and vegetable crops can be broadly classified according to their temperature requirements. However, such groupings should not be seen as absolute, because of various factors:

Unexpected limits of temperature are substantially more hurtful than more continuous ones. Defenselessness to cold or warmth may differ as per the development phase of the yield. Cabbage and lettuce are generally inclined to ice harm at the heading stage, green peas at the blossom and youthful pod stage, etc.

Cultivars of a particular vegetable contrast in their resistance to temperature limits. For instance, Winter crisp is a decent lettuce cultivar when developed during the cooler months. Be that as it may, for summer creation, Commander would be a superior decision.

Along these lines, likewise, Hercules cabbage is developed over the mid-year months, in view of its relative warmth resistance; Green Coronet has end up being a decent cultivar for winter creation. In spite of such factors, the following information can be helpful in determining probable production seasons for various vegetable crops. Reliable temperature records must be available for the production site.

3.3.2. Rainfall and Humidity

Rainfall is quite possibly the main elements, particularly when vegetables are grown under dryland conditions. Satisfactory soil dampness is vital for acceptable harvest foundation, great yields and great quality. This dampness might be gotten from rainfall or water system.

High precipitation scenes may cause flood harm, fractional suffocating on certain soil sorts, and will frequently support sickness advancement. Stickiness, or air dampness content, may likewise assume a part. High moistness will in general temper the impacts of high temperatures.

Harvests, for example, cucurbits favor dry air and a high temperature, while verdant vegetables, for example, cabbage and lettuce incline toward more muggy conditions. High dampness is more helpful for hefty dew around evening time, which can be advantageous in diminishing dampness stress, however which can support the improvement of specific illnesses, for example, leaf rust and leaf spots, on certain yields.

3.3.3. Sunshine and Day-Length

The day-length, or time of daylight every day, may affect the useful limit of vegetable yields. As an exemplary model, one could refer to the instance of onions, where taxing day cultivars will bulb just when planted beneath scopes where summer sunshine hours are sufficiently long. Cloud or fog may likewise decrease the measure of light a harvest gets, and along these lines bring down the possible yield of the harvest.

3.3.4. Wind

Wind can make huge harm from mechanical injury plants, expanded happening of plants and drying up of the dirt. On sandy soils wind-blown coarseness can make serious harm plants. Clearly, extremely breezy regions ought to be kept away from quite far, except if sufficient arrangement is made for the foundation of windbreaks.

3.3.5. Temperature and Sunlight

Maize is a crop of subtropical origin and, though it has been altered by selection for adaptation in different environments, it always responds to higher temperatures. The threshold temperature for seed germination is about 10° C.

The crop is relatively sensitive to cool temperatures, and it does not acclimatize to low temperatures as do most cool-season crops. Temperatures of 5° to 7° C may be followed by photo-inhibited physiological damage that may reduce photosynthetic rates for several days thereafter. As a C4 plant, maize responds well to both high temperatures and intense sunlight. Well-watered maize plants reach maximum leaf photosynthesis rates at midday temperatures of 32° to 35° C.

Photosynthetic rates of sun-adapted maize do not saturate until light intensity approaches full sunlight. Because photosynthetic capture of sunlight energy is the primary driving force for maize growth and yield, excessive cloudiness and short days tend to lower maize yields.

3.3.6. Precipitation and Water Requirements

Despite the fact that maize is water-efficient, the goal to acquire exceptional returns requires a lot of water. Occasional water use is around 500 to 600 mm in calm regions, and up to 900 mm or more (contingent on evaporative interest) under water system in dry environments.

Under downpour took care of conditions, which is the most widely recognized production framework, plant water is provided via occasional precipitation and put away soil water. Consequently, profound soils and those with high natural matter substance which store significantly more plant-accessible water, are considered the most reasonable for maize creation.

Water take-up continuously increments from the germination into the vegetative development stage. It's anything but a top when the harvest overhang is finished, and more specifically from not long before until soon after the fertilization period.

Estimated water use proficiency (WUE) for maize is just about as high as 2.5 kg grain per m3 of water utilized. At full overhang, water use rates might be 6 to 8 mm each day. The occasional water use by a 10-ton maize crop is ordinarily about 500mm. In the event that great maize yields are to be expected this base measure of water ought to be guaranteed.

3.4. WEATHER STRESS IN MAIZE CULTIVATION

Maize frequently suffers from weather-related problems during the growing season, the effects of which differ with the severity and duration of the stress, and the stage of crop development. Some stress conditions and their impact on crop growth and yield are discussed below.

3.4.1. Flooding

The significant pressure brought about by flooding is the absence of oxygen required for the legitimate capacity of the root framework and the lower stem. Youthful plants are killed after around 5 or 6 days of being lowered.

Passing happens all the more rapidly in blistering climate since high temperatures accelerate biochemical cycles that utilization oxygen, and warm water has less broken up oxygen. Cool climate, on the other hand, may permit plants to live for over seven days under overwhelmed conditions.

More seasoned plants at the six-to eight-leaf stage, can endure possibly more than seven days of standing water. All out submergence may, in any case, increment sickness rate and plants may experience the ill effects of diminished root development, in any event, for certain days after the water retreats.

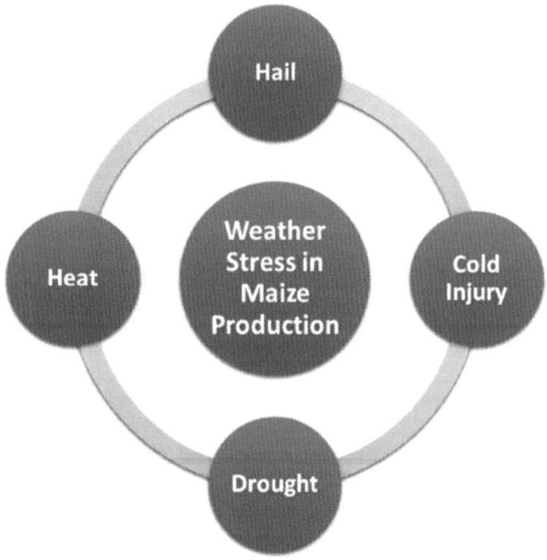

Figure 3.2. Weather stress in maize production

Resilience to flooding increments with age, yet decreased root work from absence of oxygen is more impeding to yield previously and during fertilization than during quick vegetative development or during grain fill. Submergence can likewise cover leaves with soil and diminish photosynthetic action.

3.4.2. Hail

The most widely recognized harm from hail is loss of leaf region, however tail breakage and wounding of stalks and ears can be serious also. Misfortune outlines dependent on leaf evacuation contemplates affirm that defoliation at the hour of adorning causes the best yield harm, while loss of leaf region during the primary month subsequent to planting or when the harvest is close to development by and large causes little yield misfortune. Loss of leaf region in little plants typically postpones improvement, and plants that experience hail harm may not generally develop ordinarily subsequently.

3.4.3. Cold injury

Maize isn't lenient to chilly climate. In spite of the fact that passing of leaves from ice is the clearest kind of cool injury, different tissues can likewise be harmed or killed. Loss of leaves from ice is by and large not genuine when it happens to little plants, but rather it might defer plant advancement and delay fertilization.

Ice injury side effects may show up on leaves in any event, when evening time temperatures don't fall under zero degree. Radiative warmth misfortune can bring down leaf temperatures to a few degrees underneath air temperatures on an unmistakable, quiet evening.

On the off chance that ice kills leaves before physiological development in the fall, sugars ordinarily keep on moving from the tail into the ear for quite a while, yields are for the most part lower, and collect dampness might be high because of high grain dampness at the hour of ice and moderate drying rates that typically follow unexpected passing.

3.4.4. Drought

Maize is mostly tolerant toward dry soils and dampness stress from early vegetative stages until around fourteen days before fertilization. Mild drought during mid-vegetative stages may even be valuable since roots by and large become descending all the more unequivocally as surface soils are evaporating.

The harvest additionally profits by the more prominent measure of daylight that goes with dry climate. During about fourteen days prior, during, and fourteen days following fertilization, maize is extremely delicate to dry season, and drought during this period can cause genuine yield misfortunes.

The vast majority of these misfortunes are because of disappointment of fertilization, and the most well-known reason is the disappointment of silks to arise. At the point when this occurs, silks don't get dust, and, hence, the portions are not prepared and don't create.

Developing kernels can also abort for several weeks after pollination. Drought later in grain-fill has a less serious effect on yield, though root function may decrease and kernels may not fill completely. While effects of a prolonged period of drought can never be eliminated in rain fed maize production, there are some techniques that help to reduce the effect of drought on yield. Heavy soils can be drained and tilled to depth in order to improve rooting depth.

Diminishing tillage, particularly on lighter soils, assists with preserving water and make it more accessible to the yield. Contingent upon climatic examples, changing planting date can help move the fertilization time frame into a time-frame of better water supply.

Utilizing lower plant densities, and sometimes more extensive columns, similarly give more water to singular plants and consequently stay away from complete loss of the harvest in dry regions. At last, some cutting edge maize cultivars endure or stay away from water pressure more adequately than others. Endeavors are in progress to distinguish and fuse into business cultivars qualities for dry season opposition.

3.4.5. Heat

Because dry spell and heat normally happen together, numerous individuals accept that high temperatures are a major issue for maize. The temperatures up to 40° C typically cause practically zero injury if soil dampness is sufficient. Expanded times of hot, dry breezes can cause tuft "impacting" (parching) and loss of dust suitability.

Dust shed generally happens in the cooler hours of the morning, and is frequently completed before the high evening temperatures. There is proof that crossovers differ in their affectability to both warmth and dry spell, however hereditary dry season resilience may mean some misfortune in yield potential. Subsequently, such half breeds may not be acceptable decisions for districts that typically have great developing conditions.

3.5. SOIL: THE FOUNDATION OF AGRICULTURE

Human culture has created through usage of our planet's assets in incredibly extraordinary, inventive, and useful ways that have advanced human development and supported worldwide social orders. Of these assets, soil and water have given people the capacity to create food, through horticulture, for our food.

In exploring the link between soil and agriculture, this article will highlight 1) our transition from hunter-gatherer to agrarian societies; 2) the major soil properties that contribute to fertile soils; 3) the impacts of intensive agriculture on soil degradation; and 4) the basic concepts of sustainable agriculture and soil management.

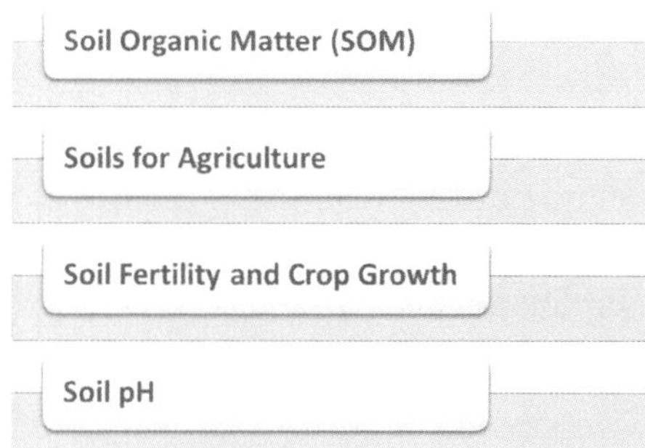

Figure 3.3. Soil: foundation of agriculture

3.5.1. Soil Fertility and Crop Growth

The early utilization of fire to flush out wild game and to clear forested land gave the main major anthropogenic effect on the climate. By consuming local vegetation, early people had the option to access herbivores brushing on the savanna and in close by forests, and to stifle the development of less attractive plant species for those simpler to scrounge and eat (Pyne 2001, Wrangham 2009).

These and different components (e.g., populace pressures, environmental change, empowering/securing attractive plants), help to lay the basis for the Agricultural Revolution and caused a sensational change in the connections among people and the earth.

The shift from agrarian social orders to an agrarian lifestyle radically shifted the direction of mankind's set of experiences and irreversibly changed regular supplement cycling inside soils. At the point when people planted the main harvest seeds at the beginning of the Neolithic Period, the dirt gave plant-fundamental supplements and filled in as the establishment for human agribusiness.

3.5.2. Plant Nutrients

Since Earth's history, normal cycling of supplements has happened from the soil to plants and animals, and afterward back to the soil, fundamentally through decay of biomass. This cycling assists with keeping up the fundamental supplements needed for plant development in the soil.

Complex nutrient cycles incorporate a range of physical, chemical, and — most importantly — biological processes to trace the fate of specific plant nutrients (e.g., N, P, C, S) in the environment. For a thorough analysis of these cycles, additional reference materials are available (Bernhard 2010, Brady & Weil 2008, Troeh & Thompson 1993).

For the most part acknowledged, there are 17 fundamental components needed for plant development (Troeh and Thompson 1993). The absence of any of these fundamental supplements, recorded in Table 3.1, can bring about a serious impediment of harvest yield — an illustration of the guideline of restricting elements.

Of the mineral components, the essential macronutrients (N, P, and K) are required in the best amounts from the dirt and are the plant supplements destined to be hard to find in rural soils. Auxiliary macronutrients are required in more modest amounts, are commonly in adequate amounts in soil, and thusly are not frequently restricting for crop development.

The micronutrients, or now and again called follow supplements, are required in limited quantities and, if in overabundance, can be poisonous to plants. Silicon (Si) and sodium (Na) are here and there viewed as fundamental plant supplements, yet because of their universal presence in soils they are never hard to come by (Epstein 1994, Subbarao et al. 2003).

Table 3.1. Essential plant nutrient elements and their primary form utilized by plants

Essential plant element		Symbol	Primary form
Non-Mineral Elements			
	Carbon	C	CO_2 (g)
	Hydrogen	H	H_2O (l), H^+
	Oxygen	O	H_2O (l), O_2(g)
Mineral Elements			
Primary Macronutri-ents	Nitrogen	N	NH_4^+, NO_3^-
	Phosphorus	P	HPO_4^{2-}, $H_2PO_4^-$
	Potassium	K	K^+

Secondary Macronutrients	Calcium	Ca	Ca2+
	Magnesium	Mg	Mg2+
	Sulfur	S	SO42-
Micronutrients	Iron	Fe	Fe3+, Fe2+
	Manganese	Mn	Mn2+
	Zinc	Zn	Zn2+
	Coppcr	Cu	Cu2+
	Boron	B	B(OH)3
	Molybdenum	Mo	MoO42-
	Chlorine	Cl	Cl-
	Nickel	Ni	Ni2+

Farming adjusts the regular cycling of supplements in soil. Serious development and gathering of harvests for human or creature utilization can viably mine the dirt of plant supplements. To keep up soil fruitfulness for adequate harvest yields, soil revisions are regularly required.

Early people before long figured out how to change their fields with creature excrement, charcoal, debris, and lime ($CaCO_3$) to improve soil richness. Today, farmers add various soil alterations to improve soil ripeness, including inorganic synthetic manures and natural wellsprings of supplements, like excrement or fertilizer, frequently bringing about surplus amounts of essential macronutrients.

The proficiency of compost application and use by crops isn't constantly improved, and abundance supplements, particularly N and P, can be moved through surface spillover or filtering from farming fields and contaminate surface-and groundwater (Moss 2008, Sharpley et al. 2002).

3.5.3. Soils for Agriculture

While soil is frequently referred to as the "fertile substrate", not all soils are suitable for growing crops. Ideal soils for agriculture are balanced in contributions from mineral components (sand: 0.05–2 mm, silt: 0.002–0.05 mm, clay: <0.002 mm), soil organic matter (SOM), air, and water.

The fair commitments of these segments take into consideration water maintenance and waste, oxygen in the root zone, supplements to work with crop dcvclopment; and they offer actual help for plants. The circulation of these soil segments in a specific soil is impacted by the five components of soil arrangement: parent material, time, environment, living beings, and geology (Jenny 1941). Every single one of these elements plays a direct and covering part in impacting the appropriateness of a soil for farming.

3.5.4. Inorganic Soil Components

As one would expect, commitments from every mineral size division help to give the actual structure to a useful soil. Loamy-finished soils are generally portrayed as medium finished with practically equivalent commitments of sand, residue, and earth. These medium-finished soils are regularly viewed as ideal for agribusiness as they are handily developed by ranchers and can be profoundly useful for crop development.

The mineral segments of soil may exist as discrete particles, yet are all the more usually connected with each other in bigger totals that give design to soil. These totals, or peds, assume a significant part in impacting the development of water and air through soil.

Sandy soils have huge pore spaces and increment water seepage, yet don't furnish soils with numerous supplements. Earth rich soils, then again, increment water holding limit and give many plant fundamental supplements.

A typical proportion of soil richness is gotten by estimating the cation trade limit (CEC). The CEC is a proportion of a dirt's capacity to trade positive particles between the dirt particles and arrangement encompassing these particles.

Because of their high surface region, earth particles can apply an enormous effect on different soil properties (e.g., CEC, structure, water-holding limit), in any event, when the percent mud content is low. Mud minerals are colloidal particles, having high surface region, with charged surfaces; allowing restricting of numerous fundamental plant supplements.

The most pervasive mud estimated particles in soils fall into the class of layer-type aluminosilicates (Sposito 2008) that ordinarily have lasting negative accuse of a high CEC. Decidedly charged dirt particles, which tie anions, incorporate those which have pH subordinate charge. The most common classes of these minerals in soils are the iron (Fe), aluminum (Al), and manganese (Mn) hydroxides (Schulze 1989).

3.5.5. Soil Organic Matter (SOM)

SOM includes the halfway or very much decayed buildups of natural biomass present in soil. SOM gives soil its profound dark tones and rich fragrances that many home landscapers and ranchers of meadow soils know about. Surface soils are made out of roughly 1 to 6% natural matter, with SOM diminishing with profundity (Brady and Weil 2002).

The 'Incomparable Plains' of North America and the 'Bread Basket' of Europe are a portion of the world's most useful agrarian soils since they created under meadow vegetation, whose root biomass and decay brought about SOM amassing. The thick dull upper layers in this dirt mirror the high SOM content.

The presence of SOM is crucial for fertile soil as it provides essential plant nutrients, beneficially influences soil structure, buffers soil pH, and improves water holding capacity and aeration. The presence of organic, ionizable functional groups (e.g., carboxyl, alcoholic/phenolic OH, enol, quinone, and amine) impart charge to SOM (Sparks 1995), contributing high CEC, and pH buffering capacity.

3.5.6. Soil pH

Frequently alluded to as the expert variable of soil, pH controls a wide scope of physical, substance, and natural cycles and properties that influence soil fruitfulness and plant development. Soil pH, which mirrors the corrosiveness level in soil, essentially impacts the accessibility of plant supplements, microbial action, and surprisingly the steadiness of soil totals.

At low pH, essential plant macronutrients (i.e., N, P, K, Ca, Mg, and S) are less bioavailable than at higher pH values near 7, and certain micronutrients (i.e., Fe, Mn, Zn) tend to become more soluble and potentially toxic to plants at low pH values (5–6) (Brady & Weil 2008). Aluminum toxicity is also a common problem for crop growth at low pH (<5.5). Typically, soil pH values from 6 to 7.5 are optimal for plant growth; nonetheless, there are some plants species that can endure — or even prefer — more acidic or essential conditions. Keeping a thin reach in soil pH is valuable to edit development. SOM and mud minerals help to support soils to keep a pH range ideal for plant development (Havlin et al. 2005). In occurrences where the pH is outside an alluring reach, the dirt pH can be changed through revisions like lime to raise the pH. Ammonium sulfate, iron sulfate, or essential sulfur can be added to soil to bring down pH.

3.6. SOIL DEGRADATION AND CROP PRODUCTION

Soil structures from new parent material through different substance and physical enduring cycles and SOM is fused into soil through decay of plant deposits and other biomass. Albeit these normal soil building measures recover the dirt, the pace of soil arrangement is exceptionally lethargic.

Hence, soil ought to be viewed as a nonrenewable asset to be saved with care for a long time into the future. The pace of soil development is difficult to decide and profoundly factor, in light of the five variables of soil arrangement. Scientists have calculated that 0.025 to 0.125 mm of soil is produced each year from natural soil forming processes (Lal 1984, Montgomery 2007, Pimentel et al. 1987, Wakatsuki & Rasyidin 1992).

In view of the time needed to create new soil, it is basic that horticultural practices use best administration practices (BMPs) to forestall soil disintegration. The dirt which is first dissolved is normally the natural and supplement advanced surface layer which is profoundly valuable for plant development.

Accordingly, the essential on location result is decreased harvest yield as just the less fruitful subsurface layers remain. Soil disintegration additionally dirties adjoining streams and streams with residue, supplements, and agrochemicals making genuine off-site impacts.

Generally, regular horticulture has sped up soil disintegration to rates that surpass that of soil development. Disintegration is frequently sped up by farming practices that leave the dirt without satisfactory plant cover and along these lines presented to raindrop sprinkle and surface spill over or wind (Singer and Munns 2006).

All through mankind's set of experiences, soil disintegration has influenced the capacity of social orders to create a satisfactory food supply.

Strong instances of this can be found in the disintegrated sediment developed in the antiquated riverbeds of Mesopotamia, making water system hazardous (Hillel 1992), and the United States Dust Bowl of the 1930s where an overwhelming dry spell expanded breeze disintegration, conveying rich dirt from the Midwest many kilometers to Washington, DC (Montgomery 2007). The Dust Bowl focused on soil disintegration in the American public cognizance of the 1930s, and it's anything but a first concern today.

3.7. SUSTAINABLE SOIL MANAGEMENT

It is apparent that, to keep up and increment food creation, endeavors to forestall soil debasement should turn into a first concern of our worldwide society. Current populace models anticipate a worldwide populace of somewhere in the range of 8 and 10 billion in the following 50 years

(Bongaarts 2009, Lutz et al. 2001) and a two-crease expansion in food interest (Alexandratos 1999, Tilman et al. 2002).

In the event that botch of soil assets keeps on decreasing the fruitfulness of the dirt and the measure of useful arable land (Pimentel et al. 1995), then, at that point we will have lost a valuable and fundamental mainstay of economical horticulture (Tilman 1999).

Practical agribusiness is a way to deal with cultivating that spotlights on production of food in a way that can be kept up with negligible corruption of environments and regular assets. This sustainable approach to agriculture strives to protect environmental resources, including soil, and provide economic profitability while maintaining social equity (Brodt et al. 2011).

The idea of sustainable agriculture is frequently misjudged to imply that compound manures and pesticides ought to never be utilized. This idea is inaccurate, as reasonable agribusiness should accept those practices that offer the most gainful types of assistance for agroecosystems and energize long haul production of food supplies in a social setting of the district.

It can't be overemphasized that maintainable practices ought to consider crop production and benefit, however should incorporate land the executives methodologies that decrease soil disintegration and ensure water assets.

By accepting certain cutting-edge innovations, demonstrated BMPs, and gaining from the past, our general public will actually want to keep on rationing soil assets and produce food supplies adequate to meet current and future populace requests.

3.8. SOILS FOR MAIZE

Maize flourishes well in many soils, the extent that they are profound (more than 1 meter) and ripe, and have a decent water holding limit. Maize develops best on sandy dirt (topsoil), loamy and silty mud soils; it is less adjusted to smaller muds and sands.

The significant soil factor is soil water stockpiling, and this is dictated by surface and construction. Accessible soil dampness is a significant component of accomplishment in downpour took care of maize creation, especially where the downpours are not consistently conveyed over the season.

For this situation the plant needs to depend on the dampness put away in the root zone to beat the transitory water shortfall. Sandy soils have a low dampness holding limit; loamy and clayey soils have a high one. Sands hold less than 5% water (50mm of water over a depth of 1 meter), while loess or clay loam soils have available water up to 20% of their volume, or 200 mm of water in one meter of soil depth. The ability of the soil to hold few weeks of water in case of dry periods during the season is a most important determinant of the potential of such soil to produce maize.

In soils where this dampness maintenance limit doesn't exist, for example in sandy as well as shallow soils, or when the moderate drought reaches out past the time that the plant can get by on put away dampness the yield will either endure or should be flooded; the last framework is frequently important for the production framework in cutting edge maize production frameworks in the western USA, Australia and other low-precipitation regions.

Maize has a generally profound root framework, coming to as much as 2 m somewhere down at times, and these roots need space to create. For ordinary root improvement, the maize crop requires a base soil profundity of 80-100 cm. Any dirt shallower than this basic profundity will give more modest yields, particularly where water system isn't rehearsed and where water will in general restrict yields.

Maize is a generally demanding harvest as far as plant supplements. It's anything but a huge amount of nitrogen (N), which is for the most part applied as a top dressing when the harvest is 25-30 cm high. The nutrients removed by a yield of 4 t grain/ha are estimated at 200kg N/ha, 80 kg P2O5/ ha and 160kg K2O/ha. In supplying the necessary nutrients for the crop, at least the same amount of these elements should be incorporated in the soil. The exact amounts of fertilizers to be added to the soil should be determined on the basis of soil analysis. Adequacy of supplies of nutrients can be assured by addition of nutrients in organic (animal or green manures) or inorganic (chemical fertilizers) forms.

3.9. SOIL REQUIREMENTS FOR GROWING CORN

Farming has to be a paying proposition — that is, the farmer must be paid a reasonable benefit as are different fragments of the economy. Until such changes occur, one approach to "beat the system" is to develop more excellent harvests with less dollar input. Harvests that order expense costs available, or when taken care of to your creatures, produce solid, high-delivering creatures. In all honesty, a significant number of our present strategies for developing yields will almost consistently deliver low quality "foodless food."

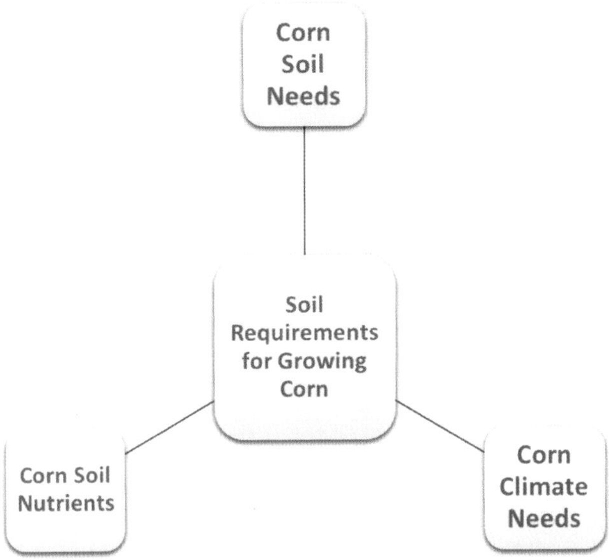

Figure 3.4. Soil requirements for growing corn

We use manures and other farming synthetic compounds that kill the life in the dirt, which whenever permitted to live would assist us with developing food. Soil turns out to be hard and tight—sterile. Weed and bug issues deteriorate.

3.9.1. Corn Soil Needs

The most fundamental thing for developing great yields is acceptable soil. Soil that isn't just high in ripeness, however is bursting at the seams with helpful life forms. The best soil for developing corn is profound (at least six feet), medium-finished and free, all around depleted, high in water-holding limit and natural matter, and ready to supply every one of the supplements the plant needs. Obviously, not every person has the perfect soil, and corn isn't particular to the point that it can't excel on not exactly ideal soil. However, I will tell you the best way to develop your dirt with the goal that you can develop much better corn.

3.9.2. Corn Climate Needs

Corn does best with warm, sunny growing weather (75–86° F), well-distributed intermittent moderate rains, or irrigation (15 or more inches during the growing season), and 130 or more frost-free days. The U.S. Corn Belt has these soil and climatic conditions.

3.9.3. Corn Humus Needs

Regardless of whether the climate isn't ideal, a decent, living soil with high humus substance will regularly have the effect between a decent harvest and fiasco, for humus permits soil to absorb extensive dampness and hold it for dry periods. It is normal the situation that one farmer who has been developing his soil will have lavish, green harvests in a dry spell year, while his neighbor's yields have caught fire.

3.9.4. Soil Parts Required For Good Corn

A normal, great soil ought to contain almost one-half mineral particles, one-fourth water, one-fourth air, and a couple of percent natural matter. The minerals supply and hold some nutrients and offer mass to the dirt. Water is essential for plant development and for the dirt life forms, yet not all that much or excessively little. Air (oxygen) is required by roots and advantageous soil organic entities.

Natural matter (humus and the living organic entities that produce it's anything but) a storage facility of specific supplements, holds water, gives soil a free brittle surface, decreases disintegration, cushions and detoxifies soil, and even shields plants from infections and nuisances due to anti-microbials and inhibitors delivered by useful microorganisms and growths.

A portion of these cordial microorganisms likewise produce plant development triggers, others help feed supplements straightforwardly to roots, and others trap (fix) nitrogen from the air—free compost. But things can go wrong. If the soil is short of air from waterlogging, low humus, compaction, or crusting, roots will suffocate or be "stunned," and the "bad guys," anaerobic bacteria, will take over and release nitrogen (denitrification) and produce several toxic substances, such as hydrogen sulfide, ammonia, aldehydes, and alcohols, when they decompose organic matter. Tight and wet soils are one of today's worst enemies of good quality crops.

3.9.5. Corn Soil Nutrients

To be healthy and produce excellent crops, a growing plant needs an adequate and balanced supply of over a dozen nutrients, mostly coming from the soil. Some are needed in larger amounts (the major nutrients:

nitrogen, phosphorus, potassium, calcium), while others are needed in smaller amounts (the secondary and trace elements: magnesium, sulfur, iron, copper, zinc, manganese, boron, molybdenum, and chlorine).

These plus carbon, hydrogen, and oxygen from the air and water are put together by the plant to form carbohydrates (sugars, starches, and cellulose), fats, proteins, vitamins, and other miscellaneous products. Photosynthesis (powered by the sun's energy) and other metabolic processes accomplish these feats.

In a living, well aerated, fertile soil, the minerals, humus, and microorganisms should supply all of the plant's needs if there are no stresses from weather.

3.10. CONCLUSION

In the coming many years, the ramifications of an unnatural weather change that is global warming in the modern economy, water supply, agribusiness, biodiversity will be exceptionally self-evident. Internationally, warming along these lines expands the recurrence and power of outrageous occasions, particularly dry season and floods.

The makes that lead the event of these marvels are clearly related both to the environment and to the human intercessions, specifically unreasonable utilization of land and water assets, inappropriate rural practices, deforestation, overburdening and, to wrap things up, air and soil contamination.

Environment information from the last century features a reformist warming of the air and a critical decrease in precipitation, which is viewed as restricting to the turn of events and efficiency of horticultural harvests, just as the utilization of water assets.

These progressions may have critical outcomes on the development and advancement of harvests during the vegetation time frame, contingent upon the level of power of the upsetting element, the mode and term of activity, and the weakness of the plant species to the event of outrageous climate occasions.

Over the long haul, measures to forestall and moderate the impacts of environmental change incorporate reforestation programs, contamination decrease, reclamation and modernization of hostile to disintegration works, just as the extension of sandy soil improvement and improvement, and so forth Simultaneously, instructing the populace and bringing issues to light about natural assurance are significant necessities in creating environmental change variation techniques.

The arrangements and suggestions for the advancement of activities and techniques to forestall and alleviate the impacts of environment changeability in horticulture should completely accept the entire arrangement of known measures (agrotechnical, social, water system, and so on) and fast mediation activities to find and restrict the spread of outrageous marvels to try not to emphasize the outcomes.

Be that as it may, tending to environmental change impacts requires specific logical information and examination, hazard the executives in horticulture, including mostly activities on the administration and protection of natural assets, just as taking right choices later on.

The establishment and advancement of rural administration choices for adjusting and diminishing the adverse consequences of conceivable environmental change on crop frameworks may suggest explicit mechanical measures: alteration of the planting date, utilization of genotypes with high protection from high temperatures/dry season, change of land use rehearses, the difference in crop pivot and, to wrap things up, the use of water system.

REFERENCES

1. 2021. CLIMATIC REQUIREMENTS. [ebook] pp.1 - 5. Available at: <https://www.kzndard.gov.za/images/Documents/Horticulture/Veg_prod/climatic_requirements.pdf> [Accessed 29 June 2021].
2. COFAS ELENA, 2018. THE DYNAMICS OF MAIZE PRODUCTION IN THE CLIMATE FACTORS VARIABILITY CONDITIONS. 9th ed. [ebook] Bucharest: The Research Institute for Agricultural Economy and Rural Development (ICEADR), pp.239-245. Available at: <https://www.econstor.eu/bitstream/10419/205114/1/ICEADR-2018-p239.pdf> [Accessed 29 June 2021].
3. E.D. Nafziger, n.d. GROWTH AND PRODUCTION OF MAIZE: MECHANIZED CULTIVATION. 1st ed. [ebook] Encyclopedia of Life Support Systems (EOLSS). Available at: <https://www.eolss.net/sample-chapters/c10/E1-05A-13-00.pdf> [Accessed 29 June 2021].
4. EcoFarming Daily, 2021. Soil Requirements for Growing Corn | EcoFarming Daily. [online] EcoFarming Daily. Available at: <https://www.ecofarmingdaily.com/grow-crops/grow-corn/soil-requirements/> [Accessed 29 June 2021].
5. Emmanuel Ofori and Nicholas Kyei-Baffour, n.d. CHAPTER 13C: AGROMETEOROLOGY AND MAIZE PRODUCTION. [ebook] Available at: <http://www.agrometeorology.org/files-folder/repository/chapter13C_gamp.pdf> [Accessed 29 June 2021].
6. Sanjai J. Parikh and Bruce R. James, 2012. Soil: The Foundation of Agriculture. Nature Education Knowledge. [online] Nature.com. Available at: <https://www.nature.com/scitable/knowledge/library/soil-the-foundation-of-agriculture-84224268/> [Accessed 29 June 2021].
7. James et al. 2000
8. OMAFRA, 2002
9. Dogan, 2003
10. Pyne, 2001
11. Wrangham, 2009
12. Bernhard, 2010
13. Brady & Weil, 2008
14. Troeh & Thompson, 1993
15. Epstein, 1994

16. Subbarao et al., 2003
17. Moss, 2008
18. Sharpley et al., 2002
19. Jenny, 1941
20. Brady and Weil, 2002
21. Sparks, 1995
22. Havlin et al., 2005
23. Lal, 1984
24. Montgomery, 2007
25. Pimentel et al., 1987
26. Wakatsuki & Rasyidin, 1992
27. Montgomery, 2007
28. Hillel, 1992
29. Alexandratos, 1999
30. Tilman et al., 2002
31. Bongaarts, 2009
32. Lutz et al., 2001
33. Brodt et al., 2011
34. Pimentel et al., 1995
35. Tilman, 1999

Chapter 4

Maize Seed Production

CONTENTS

4.1. Introduction 80
4.2. Producing Quality Seeds Means Quality Yields 83
4.3. Creating Seed-Growers' Cooperatives 84
4.4 Quality Seed Use Doubles 85
4.5. Maize Hybrid Seed Production 86
4.6. Maize Hybrid 87
4.7. Seed Classes 89
4.8. Significant Considerations for Hybrid Seed Production 94
4.9. Conclusion 98
References 100

This chapter revolves around the production of maize seed. It tells the reader about the classes of seeds which includes the pre-basic, basic, and certified seed production. It defines the seed dressing on maize. It talks about the Pannar hybrids, pioneer hybrids along with showcasing the significant considerations for hybrid seed production which incorporates isolation distance, maintenance of inbred line and seed production.

Then this chapter goes on highlighting the various other aspects such as the male-female ratio, time of sowing, the method of sowing and layout, seed rate, seed treatment. It then outlines the detasseling and drying and sorting of seed parent cobs.

It also illustrates the water and weed management, drying and shelling etc. along with defining the stages of crop inspection.

4.1. INTRODUCTION

Seed is pivotal and fundamental contribution to expand crop yields per unit region. The significance of seed in crop production is known to individual since Vedic period. There is clear notice in antiquated writing yajarveda "***May the seed reasonable, may the downpours copious and may the grains ready days and evenings***". History of agribusiness progress from early days is likewise the historical backdrop of seed of new harvests and assortments.

The advancement was quick from most recent thirty years. The green upset was just conceivable with production of by and large unadulterated seeds having different characteristics to be specific high age, high vigor high actual virtue and sound wellbeing.

Thus, green upset is seed transformation. Just seeds of guaranteed quality can be required to react to compost and different contributions to anticipated way, in any case see of expectation may transform into seed of dissatisfaction.

Among the data sources utilized by farmer, seed is the least expensive information. It is fundamental data sources and structures little piece of the complete expense of development. The great seed likewise increment the productivity of the factor of harvest creation.

Accessibility of value seeds of improved cultivars is considered essential for acknowledging usefulness and reception of cultivars in various agro-climatic conditions. The nature of seed alone is known to represent in any event 10-15% expansion in the usefulness (ICAR 1993).

Nonetheless, absence of value seed keeps on being the best obstacle so as to cross over the tremendous yield hole. In this manner, to move toward the possibly feasible yield of a cultivar, production and dissemination of value seed is fundamental. The great quality seed ought to have the accompanying characters:

- Genetic purity, and uniformity and should conform to the standards of the particular cultivar.
- Disease free, viable seeds.
- Free from admixtures of other crop seeds, weeds and inert matter.
- Acceptable uniformity with respect to size, shape and color.

Crop agriculture started over 10,000 years earlier on the flood fields of the extraordinary waterways of North Africa and the mid-East with man's disclosure of the propagative capacity of seed. From the start of harvest cultivation to our time, seeds have been the vital fixing in the foundation, extension, enhancement and improvement of yield creation.

However, seeds are a particularly indispensable piece of farming which will be underestimated by other. In other words, the people tend to ignore their vital jobs in rural production and yield improvement. Seeds are, as a matter of first importance, a generally proficient and compelling methods for crop proliferation.

They are a number of methods by which plant populaces have been and are appropriated throughout both realities. Contrasted with the vegetative propagules utilized for certain yields, seed are more modest, consequently advantageous to store and move, hardier and longer lived, simpler to plant, moderately liberated from illnesses, and a lot more modest bit of the production is needed for engendering.

While seeds are adjusted for spread of plants, they have another capacity or job which is considerably more crucial in crop improvement. Sexual cycles in plants persistently consolidate the heritable fluctuation in the between reproducing populace in various courses in the offspring created.

Seed are the result of the sexual cycle in higher plants and the vault of the heritable characteristics of the offspring. A portion of the plants engendered from seed display characteristics or blends of qualities that make them unique in relation to different individuals from the populace.

From thousands of year, the man has gradually improved the harvests he develops by saving seed for engendering from plants showing attributes that he saw to be generally valuable. In this century, the cycle of harvest improvement has been hugely sped up through precise assortment and protection of the changeability in crop species and abuse of the study of hereditary qualities.

The upgrades hereditarily designed in little populaces of plants by current plant reproducers are most effectively kept up, sent and duplicated by seeds. Seeds, accordingly, are the chief methods for expanding upgrades affected in crops, i.e., improved cultivars, from the reproducing nursery to ranchers' fields all through the space of variation.

A third, generally current, and progressively significant job of seeds in agribusiness is as a transporter of farming synthetic substances. Seeds are helpful transporters of synthetic compounds needed in generally little amounts for security of the seed and creating plants against creepy crawlies, sicknesses and weeds, controlling plant development and improvement, and correcting the supplement status of the dirt.

A decent comprehension of the base and reactant parts of seed in crop agribusiness is fundamental for the definition of compelling techniques for farming and provincial turn of events. Seed - improved seeds - should be perceived both as a vital part in the bundle of sources of info needed to improve crop creation, and as an impetus for misuse of other mechanical upgrades in crop creation.

Seeds enjoy a few upper hands over different data sources like manure, pesticides, and improved water the executives. They are needed in generally little amounts, increased as opposed to devoured in the production cycle, recognizable to all cultivars, and their utilization doesn't need any adjustment of cultivating practices fundamentally, although some changes are typically alluring.

Then again, seed when contrasted with most different information sources have two disadvantages which are frequently ignored: they are alive and should be kept up in a day-to-day environment to satisfy their propagative capacity; and the production of seed must be arranged well ahead of the time they are to be utilized - they are not a Shelf thing like compost and pesticides.

Figure 4.1. Production of maize seed

Source: Image by Flickr

4.2. PRODUCING QUALITY SEEDS MEANS QUALITY YIELDS

Farmers mainly rely upon seed confirmed by government specialists – seed that they know has been reviewed by free evaluators to convey the most ideal reap for that assortment of grain. In any case, in Central America, the seed handle, on which the smallholders rely upon, are once in a while ensured.

Governments essentially don't have the ability to investigate the gigantic number of little fields that are developing seeds of maize and beans, primary harvests of the district. Looked at side by side, it is impossible to tell the difference between fields of maize and beans being grown for food, and those being grown as seed for next year's crop. Yet there is a critical difference.

The maize and beans being developed for food are appraised for their taste or healthful substance. The ones being developed for seed are evaluated by attributes like germination reasonability, hereditary virtue and the hereditary qualities that impact their presentation in the field.

Figure 4.2. Scientist examining seeds in fields

Source: Image by Flickr

4.3. CREATING SEED-GROWERS' COOPERATIVES

The initial step was uniting farmers into seed-cultivators' cooperatives and affiliations. Having affiliations not just made it more effective to preparing, it is a unified voice to address governments' seed specialists and solicitation support.

The preparation covered creation, quality control, undertaking the board and promoting parts of seed endeavor. It consolidated the casual seed area into the conventional area by interfacing it to public agrarian examination communities and public seed specialists, while at the same time working with public governments to improve the interaction for certificate, and audit arrangements and projects that influence neighborhood markets. This all prompted the production of neighborhood organizations that presently supply improved seeds to the ranchers in their spaces.

Figure 4.3. Image of small - scale grain cooperative farmer

Sources: Image by Flickr

4.4 QUALITY SEED USE DOUBLES

The services of agribusiness and public horticultural exploration establishments in all Central American nations participated in the venture, working with FAO and its accomplices to improve seed quality by giving skill and quality control, and providing fundamental seed for smallholder seed endeavors.

As the task finished, it had multiplied the measure of excellent seed utilized in the bean and maize fields of smallholders in Central America – going from providing 8% to 16 percent of the country's seed prerequisite. Today, the ranchers' affiliations keep on expanding the quality and amount of seed they offer to their clients, who regularly are likewise their neighbors or government seed-help projects and philanthropic offices.

This progress also has had a considerable impact on food security. By December 2012, the enterprises supported by the project produced more than 6,192 metric tons of high-quality bean seed and 754 tons of quality maize seed – enough to plant 121,834 hectares of beans and 67,000 hectares of maize. But that is just the quantity.

The genuine proportion of progress is in the quality, since farmers who sowed those seeds multiplied their yields, creating enough beans to give food to in excess of 458,000 families and enough maize for more than 188,000. Planning ahead, governments' acknowledgment of the significance of this smallholder seed area has made way for proceeded with progress in approaches to help seed production and improve connections to the exploration habitats to advance continuous reproducing programs.

Today the individuals from the seed-maker affiliations keep on cooperating, zeroing in on approaches to keep expanding the nature of their seeds and, thusly, the harvest yields of the farmers who plant them.

4.5. MAIZE HYBRID SEED PRODUCTION

Maize half breed seed gives ranchers assortments containing improved hereditary qualities, for example, high return potential and extraordinary characteristic blends to counter illnesses and unfavorable developing conditions. Nonetheless, the nature of mixture seed relies incredibly upon field production strategies, both in adherence to quality affirmation principles and execution of proper agronomic administration.

While open pollinated maize seed production is generally direct, mixture seed production requires extra field practices that are basic to progress. Crossover maize seed production includes purposely crossing a female parent populace with a male parent in separated fields.

Consequently, from the actual beginning of half and half seed creation, the personality and plan of the two parent populaces decide the result. Every cross-breed assortment is made out of a particular blend of a female (seed bearing) and male (dust giving) guardians.

The field the board of the two guardians is additionally significant and expects thoughtfulness regarding timing of planting, end of off-types, expulsion of tufts from the females before dust shedding, separate gathering of the female seed and cautious shelling and preparing of the seed to keep up seed quality.

The successively reliant nature of the interaction implies that any blunders in prior stages essentially affect following stages and significant mistakes or issues can bring about complete disappointment or dismissal of the harvest.

This manual gives a manual for maize half and half seed creation, with specific reference to handle strategies, so that seed cultivators and seed organizations can cost-successfully acquire useful and quality seed appropriate for ranchers.

Figure 4.4. Image of hybrid seed production

Source: Image by Flickr

4.6. MAIZE HYBRID

Basically, a maize crossover results from the treatment of one maize plant by another hereditarily un-related plant. The plant that bears the seed is known as the female or seed parent, while the plant that gives the dust to treat the female is known as the male or dust parent.

As such, the female plant is crossed with the male plant to create half and half seed. This seed bears a one-of-a-kind hereditary make-up from the female and male guardians and will create a plant with specific attributes. Plant raisers produce the female and male guardians of every half breed to create offspring with specific attributes, for example, plant development, illness opposition, grain tone, food handling quality, etc.

It is this novel mixture seed that ranchers will plant in their fields. At the point when a farmer buys a specific half and half, the individual anticipates that the seed should act in the field as assigned by the assortment depiction.

With maize, there are various potential sorts of half and halves, like single-cross, three-way, betray and top-cross mixtures. These crossovers vary in their parental synthesis be that as it may, in all cases, the half breed seed offered to ranchers is a cross between two guardians – a female and a male.

Since maize has separate male and female plant parts, it is generally simple to make a cross between two plants. In a half and half seed production field, male and female guardians are planted in successive column designs, as a rule with three-to-multiple times the quantity of female plants or lines to a solitary male plant or line.

The male blossom (tuft) of the female plant is taken out (detasseled) before dust shed, so the lone wellspring of dust for the female bloom (the cob or ear) on the female plants is the decorations on the male plants.

Detasseling of the female is necessary to prevent any pollen from the female pollinating the female silks. If this occurs, known as "female-selfing," the result is a significant loss of seed quality that will clearly be seen in a crop grown from the seed.

Female-selfing is to be avoided at all costs. The ratio of the number of female rows to male rows in the field is usually on the order of 3:1 for single-crosses and three-ways but may extend to 8:1 for double-cross hybrids.

The real proportion that is planted relies upon various elements, however essentially on the pollen production of the male, the tallness of the male tassel comparative with the female silk and the size of the female plant. Besides, the circumstance of pollen shed of the male and silking of the female should concur.

In the event that the male and female plants are known to bloom at various occasions, change in the planting dates of every segment will be needed to guarantee blossoming synchrony of the two guardians. The half breed seed that is valuable to ranchers is reaped from the female plants.

Plants and seeds from the male lines are generally disposed of before collect to try not to blend of seed from the guardians. With this starter depiction of maize cross breed seed creation, plainly various key components decide the achievement and nature of half-breed seed creation, including the accompanying:

- Female and male parent identity, purity and identity preservation.
- Ratio of female to male rows in the seed field.
- Timing of planting of the female and male plants.

- Timely removal of the tassels from the female plants before they shed pollen and before silk emergence.
- Timing of female silk emergence relative to male pollen shed.
- Avoidance of contamination of female silks with unwanted pollen, particularly from females, off type males and foreign pollen.
- Avoidance of seed mixtures between and within the male and female plants.

4.7. SEED CLASSES

Cross breed seed production happens over various consecutively subordinate stages. The main errand is to set up the raiser's seed of the underlying innate lines. This is finished by reproducers under controlled hand fertilization to guarantee hereditary immaculateness and character.

The raiser's seed is then built up over progressive ages into adequate amounts to make the last half breed. The ages of seed building are named "seed classes" and are completed under public seed guidelines. With every age there might be a decrease in hereditary virtue, yet in the event that severe techniques are followed, this decay will be limited. Public seed guidelines determine the techniques for duplicating each seed class and contrast somewhat from one country to another, however similar standards apply, specifically:

- An identified source seed for each seed class.
- Isolation of the seed production field from potential contaminant sources, by distance (spatial) or time (temporal).
- Removal of off-type plants before flowering.
- Controlled pollination of hybrid crosses.
- Inspection of seed production field at critical stages to certify conformity to regulations.
- Identity preservation and seed labeling.

4.7.1. Pre-Basic, Basic and Certified Seed Production

The rules and procedures for certified seed production are laid down in the national seed regulations of the country in which the seed is to be produced. The aim of certification is to produce seed with an acceptable level of genetic purity and a specified seed quality in terms of minimum germination percentage (usually 90% for maize), maximum seed moisture

(12.5%) and minimum seed purity (99% pure seed with less than 3% total defects).

Genetic purity is assured by using the correct seed parents of the right seed class, developing the seed in disengagement from toxin crops eliminating off-types and, in the case of hybrids, controlling the pollination of the female. Confirmation methodology depend on guidelines for developing conditions (e.g., field history, disengagement, female-male character safeguarding, expulsion of off-types and detasseling of female plants on account of half and half creation), field assessments, predominance of weed seeds, extent of blemished seeds, germination rate and seed dampness content.

In the event that a seed field or seed parcel doesn't fulfill the endorsed guidelines for the planned seed class, it will be dismissed for confirmation. Therefore, seed makers should be comfortable with and hold fast to the public seed guidelines for the seed class that is being developed.

The principles for pre-essential and fundamental seed production are higher than those for ensured seed creation, especially as respects seclusion necessities and the greatest level of off-types. The principles are assessed in each seed field through field investigations and research facility tests.

The fields assessments are needed to confirm the beginning of source seed, recognize the assortment, decide the trimming history, check separation distance (or time) and production practices and guarantee that all accreditation methods are clung to.

Generally, three-to-five field examinations are needed through the season. Following shelling or sifting, tests of seed are taken for research center tests to assess the virtue of the seed, presence of different seeds, the germination rate and the dampness content.

The inspecting methodology guarantee that an adequately agent test is drawn from the seed part. For seed testing, the whole seed parcel ought to be available to the seed sampler to empower the assortment of a delegate test. This is the last stage in the certificate cycle, and if the seed part passes every one of the principles, it is allowed guaranteed status.

4.7.2. Seed Dressing On Maize

Most maize hybrids are delivered to producers, coated with an insecticide or fungicide, or both, as a seed dressing. Seed treatments are added to seed to protect seedlings against soil and seed borne diseases and pests, to enhance germination and to promote seedling emergence during the critical first few weeks after planting.

As a norm, maize seed is treated with a fungicide (for example Captan or Celest) and habitually likewise with a bug spray (for example Cruiser, Gaucho or Poncho), or with both, prior to planting, to shield arising seedlings from soil borne parasitic sicknesses and creepy crawly bugs. Insecticidal seed dressings can likewise shield put away maize seed from bother pervasions, for example, grain drills, weevils and grain bugs.

Some seed organizations likewise add dietary items, for example, boron and molybdenum to their seed dressings to improve seedling germination, power and rise. Unregistered or self-blended seed dressings added to seed on-ranch are profoundly questionable and isn't suggested, however are as yet being drilled by some maize makers.

Besides, the utilization of enlisted pre-rise herbicides for the powerful control of grasses and early season broadleaved weeds, are as yet rehearsed by most maize makers. Because of these practices, maize makers mentioned by means of Grain SA that the communication between various seed dressings and pre-rise herbicides be explored to decide the impacts on rise and harvest development. The partaking seed organizations have done their individual standard maize seed dressings and the examination was finished with their assent.

The cooperation impact was resolved in glasshouse preliminaries on the treated and untreated maize seed of five Pannar and six Pioneer Hi-Bred cultivars. Seed was put away at 15°C in a seed store and preliminaries were done inside 90 days after seed was gotten.

It is, nonetheless, critical to take note of that the point was not to think about half breeds from the seed organizations, but rather to take a gander at the connection of herbicides and seed medicines. The seed medicines from the two organizations were likewise not the equivalent, which make half breed correlations insignificant. Seed from Pannar was treated with Celest®, Sodium Molybdate and Cooperfos, and seed from Pioneer was treated with Captan and Poncho®.

Herbicides containing the following active ingredients: acetochlor 700 g ai (Wenner®), acetochlor 840 g ai (Guardian®) and s-metolachlor 915 g ai (Dual Gold®) were applied at the registered label rate and double the label rate. Different development boundaries like the total number of emerged seedlings, plant tallness dry mass and visual side effects of phytotoxicity, were recorded all through the perception period.

Visual phytotoxicity was seen as slant development of the coleoptile, twisting of leaves, and whiplashing of leaves, chlorosis, hindering and deformity of plant parts and was recorded by looking at sows in seed and herbicide medicines with those at the untreated control.

4.7.3. Pannar Hybrids

Crossovers with seed treatment (treated) arose inside four days in the wake of planting, while those with no seed treatment (untreated) took a normal of six days to arise. Absolute seedlings arose was higher for treated cultivars and a normal of six plants arose per cultivar, while just four plants for every untreated cultivar arose.

Phytotoxicity was generally seen as a tight collapsed coleoptile, typically dull green in shading and contorting or twisting of the spin. These side effects were noticed just a brief time after maize development. Just a single mixture showed altogether more manifestations where seed was dealt with, while the rest showed no huge collaboration among treated and untreated seed.

Seedlings, both treated and untreated seed, grew out of these phytotoxicity side effects, notwithstanding, at about a month after development of maize and no visual indications, other than hindering, was noticed ten weeks after rise of maize.

The rate visual phytotoxicity was higher for untreated seed (somewhere in the range of 17% and 30%) where herbicides were applied and were touchier to twofold herbicide measurements rates. Seedlings from treated seed showed essentially more hindering contrasted with those with no seed treatment, however hindering was under 10% where herbicides were applied at enlisted mark rates.

4.7.4. Pioneer Hybrids

Mixture seed treated with Captan and Poncho arose inside five days in the wake of planting and quicker when contrasted with seed without treatment. The untreated seed arose fundamentally later (around six days) where Wenner were applied at the name rate and twofold the name rate contrasted with treated seed and the other herbicide medicines.

Visual phytotoxicity was observed as tight folding of the coleoptile (dark green in color), twisting of the whirl and stunting, but was only observed for two weeks after emergence of maize. Untreated seed showed more phytotoxicity for all hybrids tested (between 7% and 55%).

Pioneer hybrids were less touchy to double dosage rates and just Wenner, at both dose rates, showed phytotoxicity that was industrially unsuitable. This load of cross breeds that were tried, grew out of these manifestations a month after maize development, and development was tantamount to control medicines at ten weeks after maize rise when no visual indications could be noticed.

Critical hindering was just noticed for the initial four to about a month and a half after maize development where Wenner was applied at twofold the name rate. Nonetheless, ten weeks after maize rise, all seedlings in all the herbicide medicines contrasted well and control plants and grew out of the hindering impact.

Albeit this investigation has tried a couple of mixtures from two maize seed organizations, the association between seed dressings and herbicide applications could be set up for these cross breeds assessed. Seedling rise was emphatically influenced by the different seed medicines of the significant cultivars tried and the complete number of seedlings arose were higher where seed was dealt with. Besides, all treated seed arose quicker contrasted with seed that got no seed dressing.

Stunting of seedlings was observed for both Pannar and Pioneer hybrids, but only for the first few weeks after emergence. It is, however, very important to note that all hybrids outgrew these stunting effects, as well as visual phytotoxicity symptoms ten weeks after maize emergence.

Seed treatment with Celest, sodium molybdate and Cooperfos gave extreme phytotoxic indications just as decreased dry mass. In a past report done on seed dressing and capacity conditions, it was likewise announced that Celest in mix with Cruiser, altogether diminished germination of more modest maize seed. Seeds treated with Captan and Poncho showed less phytotoxic indications and plant stature and dry mass was not antagonistically influenced by the seed dressing and herbicide communication.

Makers should observe that when some other synthetic substance(s), other than what was added by the seed organization itself, are put on seed on-ranch, it will refute any cases as per SANSOR guidelines. Horrible capacity conditions (excessively hot or cold) and unpleasant treatment of seed can likewise impact the power and ensuing germination and development of seed.

Herbicide harm to crops is seldom simply due to the herbicide(s) alone and inconvenient ecological conditions (cold and wet climate during application), decision of half breed (touchy or lenient to specific herbicides) just as incorrect measurement rates, are significant variables assuming a part in the seriousness of harvest harm.

Hybrid reactions differ greatly from each other under various environmental conditions and by adding any chemical substance(s) to the equation, crop growth is influenced in most cases by the interactions between all the factors and actions involved in the crop growth cycle.

The connections between seed dressings, hybrids, pre-development herbicides and natural conditions stay intricate and further exploration, as a team with seed organizations, is essential and will add to explain certain viewpoints saw in maize creation.

4.8. SIGNIFICANT CONSIDERATIONS FOR HYBRID SEED PRODUCTION

4.8.1. Isolation Distance

Seed production ought to be taken in rich very much depleted, weed and sickness free soil and ideally the fields where going before crop was not maize to limit rouging and keep up the hereditary virtue. Seclusion is the as a matter of first importance necessity of half and half seed production which can be accomplished through confinement by time for example fleeting segregation or through disconnection by space for example spatial disengagement. At any rate 400-500m distance is needed to stay away from any defilement. Establishment seed requires 400 m and ensured seed production requires 200 m disengagement distance.

4.8.2. Maintenance of Inbred Line and Seed Production

Support and seed production of ingrained line is a significant thought in crossover seed creation. Each innate lines are kept up in legitimate detachment just as seed of various classes are created according to endorsed guidelines. Ingrained line is an almost homozygous line created by ceaseless inbreeding, generally selfing, joined by choice. The upkeep might be finished by planting ear to push or by blending seeds from ears of individual plant of innate line for increment and might be kept up without anyone else fertilization, full sib-fertilization by hands. Typically, in numerous upkeep programs parental lines are kept up by substitute selfing and sibbing starting with one age then onto the next. For crossover seed creation, parental seeds are increased by four phase of seed chain.

Nucleus Seed------Breeder Seed------Foundation Seed-------Certified Seed.

Breeder seed and Foundation seed should never be used for commercial hybrid seed production.

4.8.3. Male: Female Ratio

The male: female ratio depends on (a) pollen shedding potential and duration of male parent; (b) male: female synchrony: for better seed setting flowering of female should be earlier than male or male pollen dehiscence should coincide with female silking and (c) season. In general, the male: female ratio should be 1:2 or 1:3 or 1:4.

4.8.4. Time of Sowing

To avoid flowering from heavy rains during fall/autumn and low/high temperature during winter season the optimum time of sowing is first week of July during fall/autumn and first week of November during winter.

4.8.5. Method of Sowing and Layout

It is desirable to plant the crop on ridges. Depending upon the plant type the row and plant spacing should be kept at 60-75 cm and 20 cm, respectively. Identification labels/ tags should be put on the male and female lines to distinguish between them.

4.8.6. Seed Rate

The seed rate depends on size of seed/ test weight, plant type and male: female ratio. 15 kg ha-1 for female and 10 kg ha-1 for male is recommended.

4.8.7. Fertilizer Application

The available quantity of the farmyard manure should be incorporated into the field before sowing. A combination of organic manure and chemical fertilizer is known to give better yield. A balanced application of 120-150kg N, 60-75 kg P,60-75kg K per hectare and 25 kg Zinc sulphate /ha is recommended. 1/3rd of total nitrogen and the entire quantity of phosphorous, potash and zinc should be applied in bands 5-7 cm deep before sowing. The rest of the nitrogen should be applied in two doses as mentioned below as top dressing.

4.8.8. Removal of Off-Type Plants and Thinning (Rouging)

- After 12-15 days of sowing: off-type and excess plants should be removed. Proper plant to plant distance should be maintained
- At knee high stage, all the dissimilar plants should be removed.
- At flowering, remove dissimilar tassel bearing plant before anthesis from the male.

4.8.9. Water Management

Water management in maize is a vital assignment since maize is a yield which can neither endure a lot of water nor do drying. Along these lines, legitimate seepage ought to be there in the seed production region and water system ought to be done just when the yield needs water at the ideal opportunity.

For the most part water management relies upon the yield planting season. Since around 80% of maize crop is filled in India particularly in downpour took care of regions. Assuming precipitation is happening on typical and right time, the harvest needn't bother with water system.

First water system should be done cautiously. Since this water system doesn't expand the development of little plants with more water. Along these lines, water in the main water system ought not to be flown over the edges.

Generally, water should be given to 2/3rd of the height of the ridge in the drains. Water requirement is high in heavy soils, whereas in the light soil the crop requires less water. The following crop stages are sensitive for irrigation in maize:

- Seedling stage
- Knee height stage
- Flowering stage
- Grain filling stage

4.8.10. Weed Management

Weed management in maize crop is a big problem and it is major problem in fall/autumn season. They compete for the nutrition, water and light in the crop which bring unexpected loss in yield. Therefore, the control of weeds is very important in order to take good yield of maize. The following weeds are mainly found in maize crop viz. Cyonodon, Dactyloctenium, Echinocloa, Cyperus, Amaranthus, Chenopodium, Parthenium, Trianthema etc.

Control: Both chemical and mechanical methods are used to destroy weeds grown in maize field.

- Atrazin: Broad spectrum pre-emergent herbicide. @1-1.5 kg a.i. /ha.
- Laudis is a broad-spectrum post emergence herbicide recommended for use along with surfactant for control of broad leaf and grassy weeds in corn
- Manual weeding

4.8.11. Detasseling

Detasseling in female ought to be done before anthesis. It ought to be practiced line astute. One individual ought to follow to screen each column to watch that no piece of the decoration is left inside. The cycle of detasseling should proceed for 8-10 days.

While detasseling, leaf ought not to be eliminated which will lessen the yield. It has been seen that the expulsion of 1 to 3 leaves alongside decoration diminishes 5-15 % yield. The eliminated decoration ought not to be tossed in the field however took care of to the cows as it is nutritive grain.

4.8.12. Harvesting

If conceivable, male parent ought to be collected after fertilization. Ideal dampness content in grain at gathering ought to associate with 20 %. The reaped cobs should spread equally as opposed to making stack.

Stages of Crop Inspection

At the time of sowing, it becomes important to monitor the land, isolation distance, planting ratio of male: female, proper sowing time, seed treatment.

- During pre-flowering/vegetative stage: to verify the rouging and removal of off type plants
- During flowering stage: to check disease and pest infestation
- During post-flowering and pre-harvest stage: to remove the late and diseased plants differential type of tassel/silk plants, harvesting time: to see the proper time of harvesting

4.8.13. Drying and Sorting of Seed Parent Cobs

The drying of the cobs should not be done either on the kuccha or pucca flour, rather it should be dried on tarpoline sheets to avoid seed injury and during night the cobs should be kept covered. To maintain the purity, dissimilar, diseased and pest infested cobs should be removed before shelling. The female cobs should be dried up to 13-14 % moisture content before shelling.

4.8.14. Shelling

Shelling of female parent should be done earlier than male to avoid mechanical mixture. Shelling can be done manually or by power operated maize Sheller.

4.8.15. Seed Processing

All under size, broken, damaged etc. seeds should be removed for maintaining the quality of hybrid seed.

4.8.16. Storage and Marketing

Seed drying should be done till the moisture content of the seed is reduced to 8 % and it should be kept in aerated jute bags. Seed should be stored at cool and dry place preferably in cold storage. Poor storage conditions will lead to loss of vigor and poor germination. Marketing should be done with specifications and standards.

4.9. CONCLUSION

Regardless of being a significant wellspring of food and feed, seeds are basically a significant conveyance arrangement of hereditary data. Seed quality is a vital deciding component of yield and nature of harvest creation. Great quality seed is better than other standard seed in hereditary and physiological immaculateness and is liberated from seed borne illnesses and problems.

The nature of the seeds is controlled by the cooperation of various hereditary and ecological elements and climatic changes essentially influence seed qualities. Production of better-quality seeds in a compelling and productive way is a test for expanding food interest.

Seed quality is an unpredictable attribute and novel examination ways to deal with improve seed quality include a mix of seed advances, hereditary qualities, and sub-atomic science. A portion of the traditional techniques for seed improvement incorporate covering, pelleting, preparing, and production of counterfeit seed.

Advancement of crossover seed assortments that adjust to troublesome climatic conditions and are impervious to a scope of nuisances and sicknesses are at the front line of the seed business in improving harvest yield.

Half breed seed assortments of rice, wheat, corn, grain, soybean, and different field crops are generally utilized in different locales of the world for improved harvest yield. Present day quality innovation techniques are being utilized to alter (GM) crops/seeds hereditarily to convey at least one advantageous attributes like herbicide and bug obstruction, better protection from dry season/waterlogging, and adjusted wholesome profiles.

Examination on hereditary qualities of seed advancement and science of seed saves is a fundamental need in growing new innovations for crop improvement. The vital test ahead is the ID and consolidation of helpful qualities and characteristics into first class cultivars, and improvement of new ways to deal with creating GM harvests to limit administrative imperatives and increment buyer acknowledgment.

REFERENCES

1. ELBÉ HUGO, 2013. Seed dressing on maize — cultivars and the interaction with pre-emergence herbicides. [online] Seed dressing on maize — cultivars and the interaction with pre-emergence herbicides. Available at: <https://www.grainsa.co.za/seed-dressing-on-maize-mdash;-cultivars-and-the-interaction-with-pre-emergence-herbicides#:~:text=As%20a%20standard%2C%20maize%20seed,fungal%20diseases%20and%20insect%20pests.> [Accessed 29 June 2021].
2. FAO, 2021. Producing quality seeds means quality yields. [online] Food and Agriculture Organization of the United Nations. Available at: <http://www.fao.org/in-action/producing-quality-seeds-means-quality-yields/en/> [Accessed 29 June 2021].
3. James C. Delouche and Howard C. Potts, n.d. THE IMPORTANCE OF SEED IN AGRICULTURE AND THE NEED FOR A SEED PROGRAM. [online] Ir.library.msstate.edu. Available at: <https://ir.library.msstate.edu/bitstream/handle/11668/13252/A-19.pdf?isAllowed=y&sequence=1> [Accessed 29 June 2021].
4. John F. MacRobert, Peter Setimela, James Gethi and Mosisa Worku Regasa, 2014. Maize Hybrid Seed Production Manual. [ebook] Mexico: International Maize and Wheat Improvement Center (CIMMYT). Available at: <https://excellenceinbreeding.org/sites/default/files/manual/98078.pdf> [Accessed 29 June 2021].
5. Mukesh Choudhary, Bahadur Singh Jat, Shankar Lal Jat and Deep Mohan Mahala, 2019. Quality Seed Production of Maize in Mizoram. [online] Research Gate. Available at: <https://www.researchgate.net/publication/331648234_Quality_Seed_Production_of_Maize_in_Mizoram> [Accessed 29 June 2021].
6. n.d. Chapter: Seed, Seed Technology and Seed Production. [ebook] Available at: <http://www.jnkvv.org/PDF/01042020205350 1.%20Seed%20and%20Seed%20Technology.pdf> [Accessed 29 June 2021].
7. Wimalasekera, R., 2015. Role of Seed Quality in Improving Crop Yields. Crop Production and Global Environmental Issues, [online] pp.153-168. Available at: <https://link.springer.com/chapter/10.1007/978-3-319-23162-4_6> [Accessed 29 June 2021].

Chapter 5

Growth and Development of Maize

CONTENTS

5.1 Introduction 102

5.2 The Maize Plant 103

5.3. Corn Growth & Development 107

5.4. Germination and Emergence 109

5.5. Reproductive Development 116

5.6. Conclusion 118

References 119

This chapter gives insight to the readers about the growth and development of maize, the maize plant and its various forms of leaves, the stalk, the roots, the tassel and the ear. It illustrates the factors due to which corn grows and develops such as the planting conditions and seed germination.

It highlights the aspects of identifying of growth of maize like the germination, emergence and it also describes the ear initiation. It highlights the significance of kernel numbers, along with explaining the silking, flowering and pollen sheds, fertilization and growth of the pollen tubes.

It also deals with the germination and emergence, different germination stages such as VE, V1-V2 stage, V3-V5 stage, V6-V7, V10-V11.

5.1 INTRODUCTION

Development of harvests, plants or plant parts is characterized as the irreversible expansion in size though advancement is the persistent change in plant structure and capacity with characteristic progress stages. Development is basically connected with capture and allocation of assets though advancement is for the most part identified with non-asset natural signals like temperature, photoperiod and light quality.

We separate turn of events and development theoretically; however the two kinds of cycles are firmly connected. Thermal time and varieties of thermal time amended to represent photoperiod and vernalization are valuable to display crop phenological advancement.

Harvest improvement, specifically the hour of blossoming, is quite possibly the main qualities for crop transformation. Reproducers, agronomists and cultivators comprehend the significance of coordinating with the example of phenological advancement to their specific surroundings, and utilize a mix of hereditary and agronomic instruments to control improvement.

Crop growth depends on the capacity of the canopy to capture CO_2 and radiation, the capacity of the root system to capture water and nutrients from soil, and the efficiency of the crop to transform resources (water, nutrients, radiation, carbon dioxide) into dry matter. Stresses such as water deficits or soil compaction reduce growth by reducing the amount of resources captured by the crop, by reducing the efficiency in the use of resources or both.

Maize is the third most significant grain crop species on the planet (after wheat and rice) and is developed across a wide scope of environments, however predominantly in the hotter calm areas and moist subtropics. Maize has various utilizations, including for human food varieties, animals feeds, and the assembling of drug and mechanical items.

It is the staple food hotspot for individuals in numerous nations. As animal feed it is profoundly attractive due to the high energy and feed worth of the piece, leaf and stem. It is getting progressively significant in numerous nations for mechanical and drug applications. It very well may be utilized to deliver starch, ethanol and plastics and as a base for anti-infection creation.

In the course of recent years, the all-out worldwide region planted to maize has expanded by about 40%, and production has multiplied. Maize is a C4 (tropical) plant. Contrasted and C3 (mild) crops, maize utilizes carbon dioxide, sun powered radiation, water and nitrogen all the more proficiently during photosynthesis.

The water-use proficiency of maize is around twofold that of C3 crops developed at similar locales. The happening proportion of maize (particles of water lost per atom of carbon dioxide fixed) is 388, while that of wheat is 613 and soybean 704.

The development and improvement of the maize plant are mind boggling measures. During the existence pattern of the plant, a significant number of the development stages cover, and keeping in mind that one piece of the plant might be fostering another part might be kicking the bucket.

Development of a maize plant is characterized as the collection of dry matter. Advancement is worried about the plant's movement from being vegetative (for example developing) to regenerative. During the existence pattern of the plant there are a few key recognizable stages at which the plant's necessities should be met to guarantee exceptional returns.

Yield supervisors should have the option to distinguish the various phases of development. The most well-known framework utilized for characterizing maize development stages separates the development cycle into two sections: vegetative growth and reproductive development.

5.2 THE MAIZE PLANT

The main structures of the plant are the coleoptile, leaves, stalk, roots, ear and tassel. Unlike all other major grain crops, the maize plant has separate male and female flowering structures – the tassel and the ear. When both flowering structures are located on the same plant, as they are in maize, the plant is called monoecious.

5.2.1. Coleoptile

The coleoptile is a pointed, altered leaf that encompasses and secures the plumule (four or five leaves moved up, one inside the other) during germination. Such assurance is significant as the coleoptile prolongs, pushing the plumule through the covering of the part (pericarp) and afterward through the dirt to the surface.

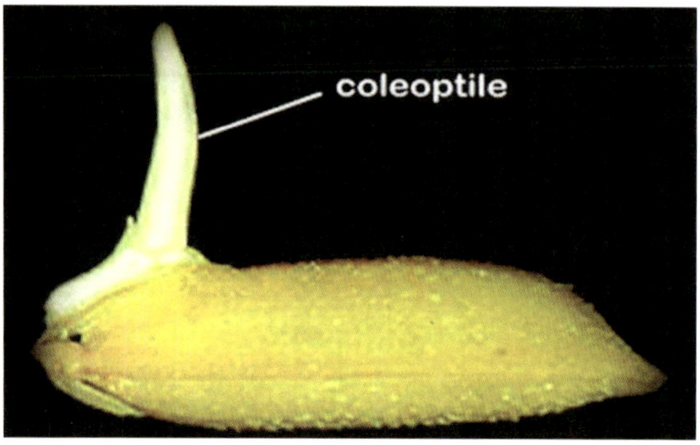

Figure 5.1. Image of coleoptile

Source:Image by Flickr

5.2.2. Leaves

The leaf comprises of the leaf sharp edge, leaf midrib and leaf collar and sheath. Leaves are created in a set request on substitute sides of the stem. The leaf edge is the almost level piece of the leaf where the cycle of photosynthesis happens.

The leaf midrib expands the length of the center of the leaf sharp edge from the base to the tip. It offers primary help. The collar is the region on the internal surface of the leaf where the leaf cutting edge and leaf sheath join. The sheath is the leaf partition (underneath the collar) that folds over a tail and joins the leaf to a tail hub.

Figure 5.2. Image of corn leaf

Sources: Image by Wikimedia

5.2.3. The Stalk

The stem is comprised of hubs and internodes and offers primary help for the leaves to capture daylight.

Figure 5.3. Image of stalk of corn

Source:Image by Pixabay

5.2.4. The Roots

The root framework comprises of sidelong fundamental roots, nodal roots and support roots. The parallel roots develop straightforwardly from the piece. These roots give safe haven to seedlings until about V3, when the nodal roots have grown adequately.

The nodal roots expand straightforwardly from every hub of the initial six to eight tail hubs underneath the ground. The support or crown establishes structure later in the season at hubs simply over the dirt surface.

5.2.5. The Ear

The ear of the maize plant is a focal cob with a circularly orchestrated gathering of blossoms, each comprising of an ovary that has a connected silk (really an exceptionally lengthened style) and is equipped for creating a portion whenever prepared effectively. On an all-around created ear there are 700 to 1000 possible portions, or ovules, organized on the cob, with ordinarily around 50 for each column long and 12 to 24 lines round.

5.2.6. The Tassel

The tassel is the male blossoming construction of the maize plant. Dissimilar to any remaining significant grain crops, the maize plant has separate male and female blossoms. The lone capacity of the male blossom (the decoration) is to deliver dust to treat the female bloom (the ear).

Figure 5.4. Image of corn tassel

Source: Image by Flickr

The number of pollen grains produced by a tassel is usually between 2 million and 5 million. If there are 1000 silks per ear, then there are 2000 to 5000 pollen grains produced per silk. Maize is a wind-pollinated plant and therefore produces copious quantities of pollen; this increases the chance of a pollen grain landing on each silk and thus the chances of fertilization occurring.

5.3. CORN GROWTH & DEVELOPMENT

In the Corn Belt, including Kansas, corn is a basic part to the accomplishment of agribusiness. All through the developing season, the corn plant goes through a progression of formative stages as it's grows from a seed at planting to a tall plant with an ear at reap. The accompanying data features different phases of development and improvement of corn all through the developing season and different issues during planting and germination.

5.3.1. Planting Conditions

It is significant to talk about seed situation at planting before coming to the plant development.

Planting shallow in the present circumstance puts the seed in a space of higher soil temperatures, which will help in quicker germination.

Planting corn into a dirt temperature going from 50 to 55 degrees Fahrenheit may take 18 to 21 days to arise, while between 60 to 65 degrees Fahrenheit can arise in 8 to 10 days. Knowing the dirt temperature at the planting profundity being utilized is significant in understanding when development can be relied upon to happen.

Another factor that is significant prior to seedling rise is buildup the executives when no-till frameworks are carried out. Imagined are two columns with differing measures of soybean buildup. The left column has a higher measure of buildup covering the line contrasted and the line on the right. Because of this, the left line has a less uniform stand.

This is because of cool soil temperatures as the buildup represses infiltration of warmth from the sun. Buildup won't just further defer development, yet can likewise cause lopsided stands, which can likewise diminish yields. The board of corn or sorghum buildup is more basic because of the greater sums that stay in the field contrasted with soybeans.

Soil dampness additionally requires thought at planting. For this situation, this seed was planted when the dirt was excessively wet, which caused crusting. Crusting happens when water and pressing factor causes soil particles to "stick" together, making a layer that is hard to enter.

In case of crusting happening in the field, precipitation is typically the best strategy to tackling the issue. Nonetheless, precipitation can likewise cause crusting to happen if the precipitation power is excessively hard.

5.3.2. Seed Germination

Since the seed is in the soil, it will start to retain water and start to expand. At the point when enough water has been consumed and the dirt temperature is good, germination will happen. The main root called the radicle will then, at that point arises. The image represents the radicle starting to stretch, around ten days subsequent to planting.

Soon after emergence of the radicle, the mesocotyl will begin to elongate. The coleoptile, which is a protective sheath over the mesocotyl, is seen first when emergence occurs. Mesocotyl elongation is very sensitive to soil temperatures. If soil temperatures are too cool, elongation will be slow and emergence will be delayed. Pictured are the developing mesocotyl and coleoptile.

Figure 5.5. Image of seed germination of maize

Sources: Image by Wikimedia

5.2.3. Identifying Stages of Growth

Corn development stages are classified as vegetative (V) or regenerative (R). The leaf collar strategy is one technique used to organize corn while it is in vegetative stages. With this strategy, the vegetative stages depend on the quantity of apparent leaf collars.

Organizing starts at rise (VE), and each new leaf with a completely evolved leaf collar is called stage V (n). Vegetative leaf arranging for corn plants closes when the corn fosters a decoration (VT); when the plant has silks noticeable outside of the husks it has arrived at the main conceptive stage (R1).

Vegetative and conceptive stages are resolved on an entire field premise when at any rate half or a greater amount of the plants have reached or are past a specific stage.

5.4. GERMINATION AND EMERGENCE

5.4.1. Germination – VE

At the point when the maize kernel is planted in damp, warm soil, water is consumed through the seed coat (this cycle is called imbibition) and the bit starts to grow. The bit collects about 30% of its dry load as water before enzymatic movement in the undeveloped organism starts.

Compounds stimulate development in the undeveloped organism, and if conditions keep on being positive the radicle lengthens and rises out of the seed coat inside 2 or 3 days. The plumule likewise starts to extend, and extra leaves start to frame inside this piece of the creating seedling.

The minuscule, delicate leaves are encased in a specific leaf structure called the coleoptile, which is pointed on the end and ready to develop vertically through rough soil to carry the departs to the surface.

The principal seedling root (the radicle) is before long followed by a few other fundamental or seed roots. These anchor the creating seedling and assume a part in water and supplement take-up. The roots that emerge from the portion are known as the essential root framework.

The primary roots (seminal roots) grow directly from the kernel and maintain the seedling for the first 2 to 3 weeks until the permanent (secondary) root system develops from the crown of the plant. The crown contains the growing point of the seedling and remains about 3 cm below the soil surface until about 3 weeks after emergence.

Between the place of connection to the kernel and the crown is the mesocotyl, a rounded, white, stemlike part here and there viewed as the principal internode. Both the coleoptile and the mesocotyl stretch as the tip of the coleoptile advances toward the dirt surface. Once the coleoptile breaks the soil surface and experiences daylight, it quits developing and the principal leaf gets through the coleoptile.

At the point when light strikes the coleoptile tip, a compound sign delivered from the arose tip assists with fixing the profundity of the crown, generally around 2 to 3 centimeters underneath the dirt surface. The profundity of the bit is subsequently not as significant as this sign in deciding the profundity of the crown.

The length of the mesocotyl is ordinarily equivalent to about a large portion of the planting profundity, however in the event that the part is planted profoundly, the mesocotyl can be any longer, since the profundity of the crown is pretty much steady.

5.4.2. Emergence – VE

The coleoptile rises out of the dirt 6 to 10 days subsequent to planting, yet this can be deferred by cool temperatures or dry soil. When the coleoptile tip arrives at the daylight, it parts at the tip and two genuine leaves unfurl.

In cloddy soil or within the sight of dirt outside layer or certain herbicides, the coleoptile tip can crack underneath the dirt surface, rashly delivering the primary leaf. Large numbers of these plants kick the bucket, as the leaves are not sufficiently able to push through the dirt.

By 7 days after rise, the new seedling has two completely extended leaves. The essential root framework has sufficiently grown to make the plant free of the diminishing food supply in the part. After the initial two leaves arise, the following leaves develop vertically inside these leaves (for example up through the 'whorl') and unfurl at the pace of around one leaf at regular intervals, or around one leaf for each 65 developing degree days that collect.

5.4.3. Vegetative Growth

V1–V2 stage

These growth stages occur about 1 week after the plant emerges. At this stage 1 to 2 leaves are visible on the stem and the primary root system is still relatively small. The plant is using only small amounts of nutrients but it needs to be in contact with the root system. Banded fertilizer placed underneath the seed will allow effective uptake.

V3–V5 stage

Fourteen days after the plant arises, the V3 stage starts. The seedling root framework has quit developing and the nodal roots presently structure the significant piece of the root framework. Root hairs are available on nodal roots. The foundations of the subsequent whorl are stretching. Development too close to the plant will annihilate a portion of the lasting root framework.

Leaf and ear shoots are being initiated, and this initiation will be complete by V5 (when the potential ear shoot number is determined). Also, by V5, a microscopically small tassel is initiated at the growing point. The plant is around 20 cm high when the tassel is started, yet the developing point is still at, or simply under, the dirt surface.

Be that as it may, soil temperature can influence the developing point. Cold soil temperatures at this stage can decrease supplement accessibility and moderate formative stages. Ice or hail may annihilate the uncovered leaves yet won't harm the developing point beneath the dirt surface. Leaf harm as of now for the most part brings about almost no decrease in conclusive yield.

Contest with weeds for water, supplements and light at this stage, weeds seek water, supplements and light. Synthetic substances, development and higher plant populaces or yield turn utilized in crop arranging can decrease weed pressing factor and breaking point the opposition to the maize crop.

Ear Initiation

The ear, or female blossoming structure, is a focal cob with a roundly showed gathering of female blossoms. Each bloom comprises of an ovary with a joined silk (an exceptionally prolonged style) and is fit for delivering a piece whenever prepared effectively.

On a well-developed ear, there are 700 to 1000 potential kernels or ovules arranged on the cob, with about 35 per row and 12 to 24 rows. The timing of ear initiation varies with maize genetics. An overall rule is to figure out which hub contains the essential ear (the ear to be collected) and afterward deduct seven to give the stage number.

This will give a reference point concerning when ear commencement began. The V stage when ear commencement began is likewise when the quantity of bit columns around the ear was being set up. This would imply that ear inception started at or close to the V7 stage. The area of the essential ear additionally differs with maize hereditary qualities.

The essential ear is situated on the V14 hub. As a rule, maize lines with a relative development between 103 to 118 will create the essential ear on the V13 or V14 hub. Parent lines of prior development will put the essential ear on a lower hub, like the V12 hub, though maize lines of longer development may put the essential ear on a higher hub.

Establishing the Number of Kernel Rows Around the Ear (V5–V8 Stage)

Contingent upon its relative development, the maize plant decides the greatest number of columns around the ear between the V5 to V8 stage in its life cycle. The meristematic arch at the tip of the ear shows that the creating ear is as yet delivering new lines of ovules along its length. The upper two thirds of the ear show a progression of single columns of creating ovules.

These ovules in the long run separation to deliver a couple of lines of ovules from each single column. This matched arrangement is apparent close to the foundation of the ear. This division into sets clarifies why a maize ear consistently has a much number of kernels columns around the ear. The most extreme number of ovules that the whole maize ear will create will be resolved after around four more V stages.

Establishment of the number of kernel rows around the ear is a critical event and is susceptible to environmental stress. Knowing when this event occurs helps to establish a 'time window' when looking for a reason. For example, if an ear has 12 kernel rows around instead of the normal 16, then the stress factor that caused this event was present at approximately V7.

V6, V7

Approximately 3 weeks after the plant emerges, the plant is at the V6 stage. The root system is well distributed in the soil and extends about 45 cm deep and 60 cm wide. The third root whorl is elongating. The growing point is above the soil surface and rapid stem elongation begins.

The plant is presently retaining more noteworthy measures of supplements. Column applied manure is less basic now, as nodal roots have spread all through the dirt. Nitrogen can be side-spruced up to V8 in the event that it is set in clammy soil and overabundance root pruning and injury of over-the-ground plant parts are stayed away from. This is a basic stage for yield when the maize ear is setting the most extreme number of piece columns around the ear.

Setting Up Kernel Numbers (V7 – Pollination)

From about V7 until pollination the maize plant is deciding the length of the ear and the quantity of kernels along the ear. Ovules close to the foundation of the ear foster first, and more current ovules will keep on shaping as improvement advances towards the tip of the ear.

After the maize plant has established the maximum number of ovules, the energy, nutrients and water to sustain these developing ovules must be supplied. If resources are adequate, ovules along the entire ear will develop sufficiently to produce silks and be receptive to pollen. If resources are limited, selected ovules will be sacrificed to permit the maize plant to satisfactorily uphold the excess feasible ovules. These ovules are relies upon the sum, type and span of the pressure. The ovules close to the tip of the ear are forfeited, bringing about suitable ovules just at the foundation of the creating ear.

Ovules close to the foundation of the ear are bound to stay practical on the grounds that these ovules are more evolved and are nearer to the wellspring of supplement supply. Provided with the fact that the ecological pressure is short however extraordinary, the ovules that are forfeited might be anyplace along the maize ear.

Exceptionally short ears, called 'brew can' ears have all the earmarks of being because of a blend of natural burdens – conceivably chilly pressure combined with dry spell pressure during a basic stage in ovule development.

V8, V9

Around a month after the plant arises, it is at the V8 stage. The fourth whorl of nodal roots is prolonging. A few ear shoots are available. A potential ear shoot will frame at each over the ground hub aside from the upper six to eight. At first, every ear shoot grows quicker than the one above, however development of the lower ear shoots eases back.

Just the upper a couple of ear shoots ultimately structure harvestable ears. Productive crossovers will in general shape more than one harvestable ear, particularly at lower plant populaces. Supplement insufficiencies at this stage will confine leaf development.

Removal of all the unfurled leaves of the plant at this stage (by frost or hail) may result in 10% to 20% reduction in final grain yield. Water logging at this or any earlier stage, when the growing point is below ground, can kill the maize plants in a few days, especially if temperatures are high.

V10, V11

Around 5 weeks after the plant arises the V10 stage starts. The maize plant quickly gathers the two supplements and dry matter. The time between the presence of new leaves is abbreviated, with another leaf currently showing up each 2 or 3 days. Interest for soil supplements and water is high to address the issues of the expanded development rate.

Dampness and supplement insufficiencies at this stage will particularly impact the development and improvement of the ears. Manure is required close to the roots – particularly phosphorus and potassium, which don't move in many soils.

Two yield components – the potential number of kernels and the ear size – are being determined during the period from V10 to V17. The length of time for the plant to develop through stages V10 to V17 affects harvestable yield.

Early-developing hybrids regularly progress through these stages in less time and have more modest ears than later maturing half and halves. Thusly, higher plant populaces are required for before cross breeds to create grain yields like those of mid-to late-development crossovers.

V12, V13

Roughly a month and a half after the plant arises, the V12 stage starts. Support roots are creating from the fifth hub and the first over the ground hub. Development of plants right now will annihilate a portion of the plant roots. By the V12 stage, the potential number of kernel rows and the potential number of ovules is established.

The meristematic arch is not at this point present, so ovule arrangement is currently finished. Combined ovule development is obvious along almost the whole length of the ear. On the off chance that an ear has the legitimate number of portion lines around the ear however the ear is more limited than ordinary, then, at that point pressure or some likeness thereof while the maize plant was around the V12 formative stage may have caused this occasion.

V14, V15

Around 7 weeks after the plant arises, V14 starts. The maize plant at V15 is simply 12 to 15 days (around five V stages) away from the beginning of conceptive development (R1, 'silking'). This vegetative stage is the most basic time of bit yield assurance. The quantity of ovules that foster silks, and hence the quantity of bits, is being resolved.

Any injury or supplement or dampness lack (like hail or creepy crawlies) may genuinely diminish the quantity of pieces that create. The decoration is close to full estimate yet not noticeable from the highest point of the leaf sheaths. Silks are simply starting to develop from the upper ears. Upper ear shoot improvement has outperformed that of lower ear shoots.

Another leaf stage can occur each 1 or 2 days. Support roots from the 6th leaf hub are creating and the perpetual roots have proceeded to extend and multiply, ultimately arriving at a profundity of around 100 cm and spreading every which way. In certain half and halves, support roots will likewise create from the eighth and ninth leaf hubs or much higher.

V16, V17

Approximately 8 weeks after the plant emerges, it is entering the late vegetative stages if it has not already developed its total number of plant leaves. During this time, plant stress can greatly affect yield. Moisture stress 2 weeks before or after silking can cause a large grain yield reduction.

This is additionally valid for different sorts of natural anxieties (hail, high temperature, supplement lacks). The 4-week term revolved around silking is the best an ideal opportunity for water system if water supply is short.

The tips of the upper ear shoots might be apparent at the highest point of the leaf sheaths by V17 in mixtures that foster in excess of 16 leaves. The tip of the decoration may likewise be apparent by V17 in more productively leafing half and halves.

V18 or VT

The vegetative plant is arriving at full size in productive leafing mixtures. Silks from the basal ear ovules have been quick to lengthen, trailed by the silks from the ear tip ovules. Support roots are presently developing from over the ground hubs. These support attaches offer help to the plant and acquire water and supplements from the upper soil layers during the conceptive plant stages. Ear improvement is proceeding quickly, with the plant just multi week from reasonable silking. Stress in these later vegetative stages can defer the start of dust shed. In the event that dust shed is postponed, plants may defer silking until pollen shed is part of the way or totally wrapped up. In the event that ovules are unfertilized there will be missing bits on the ear, particularly at the ear tip.

5.5. REPRODUCTIVE DEVELOPMENT

5.5.1. Tasseling – VT

The tassel is the male blossoming design of the maize plant. The lone capacity of the tassel is to deliver sufficient dust to treat the ovules in the female bloom (the ear). A lively maize decoration can create between 2 million and 5 million dust grains.

In the event that there are 1000 silks for every cob, there are 2000 to 5000 dust grains delivered for every silk. An enormous volume of dust is required in light of the fact that the male and female blossoms are actually isolated and there are countless florets on each plant. A huge volume of dust guarantees there is in every case sufficient dust for each uncovered silk.

Regularly, the tip of the tuft can be seen at about the very time that the tip of the arising ear gets apparent. The decoration rises out of the encasing leaves before dust shed starts, which is normally 1 or 2 days before silks initially show up.

5.5.2. Flowering (Pollen Shed and Silking) – R1

There are two parts to the pollination process:

- Viable pollen must be shed and land on receptive silks
- The pollen must germinate and form pollen tubes to allow male gametes (sex cells) to fuse with female gametes inside the ovule.

5.5.3. Pollen Shed

Dust grains create and develop in anthers of the tuft, every one of which contains countless dust grains. The anthers rise up out of the encasing glumes, typically ahead of schedule to early in the day after the dew has dried from the tuft. The anthers split open and shed dust into the air. At least 100 dust grains/cm2/day are expected to effectively fertilize a maize field.

Dust grains are light and can be carried on the breeze, however most settle inside 5 to 15 m of where they were delivered. One full-shedding decoration can give sufficient dust to a few years, if the circumstance and dispersion of dust are correct.

Dust is shed for a few days (regularly 5 to 10), with top production on about the third day. Dust shed starts from the center of the focal spike of the tuft, fans out over the entire decoration in succeeding days, and finishes with shed from the tips and bases of the lower branches.

Dust shed is definitely not a constant cycle. It stops when the decoration is excessively wet or excessively dry and starts again when temperature and dampness conditions are positive, or when extra dust has developed.

On a typical clear mid-summer day, peak shedding occurs between 9:00 and 11:00 am (Australian Eastern Summer Time). Despite the overlap in timing of pollen shed and silking, pollen of a given plant rarely lands on silks of the same plant.

This is because of air development, leaf arrangement, and the upward division between the decoration and the ear. Under typical field conditions, as numerous as 97% of the bits delivered by a plant create from dust shed by different plants in the field.

5.5.4. Silking

Normally, the main silks delivered on a plant rise up out of the encasing husks 1 to 3 days after dust shed has started. Under favorable growing conditions, all silks will emerge and be ready for pollination within 3 to 5 days – enough time for all silks to receive pollen before the tassel stops shedding pollen.

Just like the case with other plant parts, silk development happens for the most part around evening time; examines have shown that silks can grow 5 to 8 cm in a solitary evening.

The silks from close to the foundation of the ear arise first, subsequent to developing for a few days, and those from the tip show up last. Silks by and large show up external the husks toward the finish of the ear in the first part of the day, prepared to get dust as it is shed. When viable pollen grains fall on maize silks, they are trapped by small hair-like projections called trichomes.

5.5.5. Fertilization and Growth of The Pollen Tubes

Dust grains contain starch as an energy supply, and they grow inside a couple of moments subsequent to arriving on the trichomes. They produce a dust tube that develops down the silk channel and enters the ovary. These dust tubes appear to consistently develop close to the silk. This is on the grounds that the silks supply the entirety of the important supplements and water for development of the dust tubes.

The dust tube develops the length of the silk in 12 to 28 hours. The dust tube breaks at the tip to deliver cores into the ovule, treating both the egg, which forms into the incipient organism, and the polar cores, which form into the endosperm of the new kernel.

After fertilization, an abscission layer forms near the kernel and the silk detaches, leaving a tiny 'silk scar' on the kernel. Silks that are not pollinated continue to grow, sometimes reaching a length of 15 cm or more. Contingent upon water accessibility and natural conditions, it might require only a couple hours to around one day for pollen tubes to develop the entirety of the path to the ovules. At the point when the maize plant is under more noteworthy dampness stress, dust tube development is increasingly slow potential for fruitful treatment diminishes.

5.6. CONCLUSION

The development pace of a corn plant is delayed toward the start of the period, yet increments with the presence of each new leaf. Under non-upsetting conditions, the time between new leaves will diminish as the season advances.

The plant is generally powerless against pressure during silking, when significant fertilization occasions are happening. As the regenerative stages progress, the impact of weight on seed weight will diminish, while the impact on seed number will be insignificant after R2.

Most significant returns will be accomplished in regions where ecological conditions are great for these development stages, particularly R1. Negative conditions from the get-go in the season will restrict leaf size, which will diminish photosynthesis, while stress later in the season can influence pollination as kernel size and number.

REFERENCES

1. Dr. Kraig Roozeboom, n.d. Corn Growth & Development | Corn | Crop Production | Agronomy | Kansas State University. [online] Agronomy.k-state.edu. Available at: <https://www.agronomy.k-state.edu/extension/crop-production/corn/corn-growth-and-development.html> [Accessed 29 June 2021].
2. Dr. Mark Licht, 2021. Corn Growth Stages | Integrated Crop Management. [online] Crops.extension.iastate.edu. Available at: <https://crops.extension.iastate.edu/encyclopedia/corn-growth-stages> [Accessed 29 June 2021].
3. Jan Edwards, 2009. Maize growth & development. [ebook] State of New South Wales through NSW Department of Primary Industries. Available at: <http://ISBN 978 0 7347 1955 3> [Accessed 29 June 2021].
4. Sadras, V., Villalobos, F. and Fereres, E., 2016. Crop Development and Growth. Principles of Agronomy for Sustainable Agriculture, [online] pp.141-158. Available at: <https://link.springer.com/chapter/10.1007/978-3-319-46116-8_11> [Accessed 29 June 2021].

Chapter 6

Weed Management in Maize

CONTENTS

6.1. Introduction 122
6.2. Weeds 125
6.3. Major Weed Species in Association with Maize 128
6.4. The Critical Period of Crop-Weed Competition 130
6.5. Weed Management Program in Maize 133
6.6. Effect of Pre-Sowing Herbicides on Weed Control of Maize 138
6.7. Impact of Pre-Rise Herbicides on Weed Control in Maize 139
6.8. Effect of Post-Emergence Herbicides on Weed Control in Maize 140
6.9. Integrated Weed Management in Corn 141
6.10. Conclusion 142
References 143

This chapter gives an insight to the reader about how weed is managed and organized in maize. It explains the production of maize in Eastern Gangetic plains. It further talks about the weed, its characterization along with its advantages and disadvantages. It illustrates the major species of weed in association of maize.

In the second half, this chapter puts emphasis on the critical period of crop weed competition, the control in production of weed by chemical measures. It goes on defining the weed management programs like the socializing, mechanical process of weed control, and the chemical weed control like the flame weeding as well as the natural sources.

It then talks about the pre-rise herbicides on weed control in maize along with outlining the post- emergence herbicides on weed control in maize. It also highlights the incorporated weed management in corn.

6.1. INTRODUCTION

Weeds are plants that are undesirable in a given circumstance and might be destructive, hazardous or financially negative. Weeds are a genuine danger to essential production and biodiversity. They decrease homestead and timberland efficiency, uproot local species and contribute essentially to land and water corruption.

The expenses of weeds to the regular habitat are additionally high, with weed attack being positioned second just to natural surroundings misfortune in causing biodiversity decay. In spite of extensive government and private area speculation, weed intrusion actually addresses a significant danger to both the useful limit of land and water and the uprightness of our regular environments.

A proficient weed control program must be created after the weed has been appropriately distinguished. Weeds can be overseen utilizing various techniques. The best administration of weeds is normally accomplished through joint effort and collaboration, in associations between the local area, land proprietors, horticulture, industry and the different degrees of government, utilizing a mix of strategies related to an intensive subsequent mission.

Weed the board is a significant segment of plant security improving the production capability of harvests. It remembers the board of the weeds for a way that the yield supports its production potential without being hurt by the weeds. Weed the executives is done through the mechanical, social and synthetic methods. Utilization of organic control strategies in field crops is being thought of, yet very little being used. Utilization of herbicides is a significant technique in the cutting-edge idea of much being used. Utilization of herbicides is a significant strategy in the cutting-edge idea of weed-the executives innovation. New hand-devices and executes have likewise been intended to aid marry the board program.

Weed control is fundamental to accomplish ideal maize yields. The main significant period for weed control is from planting to about a month and a half of development. At planting, it is suggested that the seedbed should be without weed.

The maize will for the most part sprout following five to seven days, during which time the seedlings are entirely powerless against weed contest. Weeds rival the maize for supplements, water and space. They are more adjusted to rivalry and, if not controlled, the maize yield will decay.

Weed the executives predominantly centers on yearly weeds than lasting weeds. Annuals complete their lifecycle from one seed to another in under one year or in one developing season. They produce a plenitude of seeds, develop rapidly and are ordinarily, yet not generally, simple to control, for example pigweed, normal blackjack and gallant soldier.

Perennials are divided into two groups, creepers and simple perennials. Simple perennials spread only by seed. If the shoot is injured or cut off, it may generate a new plant vegetative, but reproduction is normally done by seeds.

Creeping perennials reproduce by seed and vegetative plant parts. These include above-ground stems (stolons), below-ground stems (rhizomes), tubers, aerial bulblets and bulbs, e.g., yellow nuts edge.

Make use of an integrated weed management programme, combining cultural, mechanical and chemical methods.

Eastern Gangetic Plains (EGP) covering three adjoining nations (India, Bangladesh and Nepal) has an extremely high return capability of maize creation. In late many years, maize an arising as a money crop for smallholder ranchers in view of its high return possibility because of great climatic conditions which permits maize production round the years in three seasons. The EGP is essentially overwhelmed by rice-based cultivating frameworks having greatest inclusion of rice-rice cropping framework which is related with own issues of high-water utilization, production expenses and work use, and further soil wellbeing disintegration.

Figure 6.1. An illustration of weed plant

Source: Image by Wikimedia commons

This load of related issues makes rice-rice production unreasonable and unbeneficial. Ranchers are searching for the elective harvest of dry season rice called boro rice (winter rice) to differentiate their trimming frameworks and maize is perhaps the most appropriate harvests to give high return and net revenue in a more reasonable manner.

As of now, over 90% maize fills in EGP by manual drilling after intensive tillage operations (4–6 dry tillage) which postponed the maize planting by at the rate of 1-2 weeks. Various dull tillage tasks increment the expense of development because of higher energy and work use and deferred planting which further decreases the net edges of the ranchers.

Since, the preservation rural (CA) based administration practices (least/ zero culturing) are the new arising innovations growing a consideration worldwide because of higher monetary advantages, improvement in soil wellbeing and furthermore discovered more ecological amicable by decreasing ozone harming substances.

In spite of the fact that CA-based maize development has numerous advantages, anyway weed pervasion is one of the genuine requirements that limits maize production however weed the executives is likewise a significant worry in customary production framework.

The opposition among weeds and maize at basic development stages could be diminished both the quality and amount of maize yield more than 30% as weeds contend with crop for fundamental assets. For controlling weeds in crop field, ranchers are for the most part embracing mechanical, social, natural and synthetic control techniques.

Among them, depleted by social techniques, ranchers are moving towards other elective strategies because of work emergency during basic time of weed control. While, the mechanical strategies are as yet helpful however can't viably control of weeds effectively because of the shortfall of right hardware.

The wise and right utilization of an alternate blend of herbicides as pre-plant, pre-rise and post-development can give compelling and effective weed control under CA framework just as traditional framework. The synthetic weed control might be given practical, quicker and better weed control.

6.2. WEEDS

Any plant not planted in the field by the farmer and develop where they are not needed is known as a weed. Weeds resemble some other harvest plants in size, structure, morphological and physiological characters however have the accompanying attributes by virtue of which they are considered as foe of yields.

6.2.1. Characteristics of Weeds

- The weeds seed germinate early and the seedlings grow faster. They being hardy compete with the crop plants and deprive them of light, moisture and nutrients.
- They flower earlier, run to seed in profusion and mature ahead of the crop. They are, therefore, difficult to control and it may be even impossible to eradicate some weeds completely.
- They are unusual, unwanted and undesirable.
- They are harmful to crops, cattle and human-beings.
- They can survive even under adverse conditions.
- They are prolific and have a very high reproductive capacity.
- Viability of seeds remains intact, even if they are buried deep in the soil.

- The seeds may have special structures like wings, spines, hooks, sticky hairs etc. on account of which they can be easily disseminated over long distance.
- Many weeds like *doob* are vegetatively propagated and spread rapidly all over the field even under adverse conditions.

Figure 6.2. Weed are found to be harmful to crops, cattle and humans as well

Source: Image by Pixabay

6.2.2. Harmful Effects

Weeds truly affect horticultural creation. It is assessed that overall weeds cause 5% misfortune in agrarian production in the majority of created nations, 10% misfortune in less created nations and 25% misfortune in least created nations. In India, yield misfortunes because of weeds are more than those from bug and infections.

Yield misfortunes because of weeds shift with the harvests. Each harvest is presented to serious rivalry from weeds. A large portion of these weeds are self-planted and they give rivalry brought about by their quicker pace of development in the underlying phases of yield development. In certain harvests, the yields are decreased by over half because of weed pervasion.

Weeds contend with crops for water soil, supplements, light, and space, and accordingly diminish the harvest yields. A gauge shows that weeds can deny the yields 47% N, 42% P, half K, 39% Ca and 24% Mg of their supplement take-up. Weeds also act as alternate hosts that harbor insects, pests and diseases and other micro-organisms. They are also alternate hosts of some of the pest and diseases.

Some weeds release into the soil inhibitors of poisonous substances that may be harmful to the crop plants, human beings and livestock. Health problems caused by weeds to humans,

Weeds reduce the quality of marketable agricultural produce. Contamination of weed seeds of Datura, Argemone, Brassica etc., is harmful to human health and weed seeds present in the produce cause odd odor sometimes.

Weeds decrease yield as well as meddle with agrarian activities. Weeds make mechanical planting a troublesome cycle and render reaping troublesome, prompting expanded use on work, gear and synthetic substances for their evacuation.

In sea-going climate, weeds block the progression of water in trenches, water-transport framework and waste framework, delivering route troublesome. The thick development of amphibian weeds contaminates water by deoxygenating it and killing the fishes. Weeds are also a pest and a fire hazard along railway lines, roads, right-of- ways, airports, forest and industrial sites.

6.2.3. Beneficial Effects

Disregarding every one of the troubles brought about by weeds, they can offer some valuable properties, especially while happening at low densities. These perspectives ought to be used in the cultivating framework, albeit this may make natural administration more muddled than substance-based frameworks. A portion of the likely advantages of weeds are provided below:

- Helping to preserve soil dampness and forestall disintegration. A ground front of weeds will diminish the measure of uncovered soil presented assisting with rationing supplements, especially nitrogen which could somehow be filtered away, particularly on light soils.

Figure 6.3. There are various benefits of weed in medical science

Source: Image by Wikimedia commons

- Food and sanctuary can be accommodated regular foes of irritations and surprisingly elective food hotspots for crop bothers. The real presence of weed cover might be a factor in expanding adequacy of natural control of irritations and decreasing vermin harm.
- Weeds can also be valuable indicators of growing conditions in a field, for example of water levels, compaction and pH.
- Weeds can be an important source of food for wildlife, especially birds. Bird populations have been declining on farmland over the last few decades and leaving weeds as a resource has been shown to help revive bird populations.

6.3. MAJOR WEED SPECIES IN ASSOCIATION WITH MAIZE

Swarming weed species in crop field by and large rely upon crop species, trimming frameworks and their administration practices, ecological conditions, occasional variety, soil properties, supplements and dampness status just as soil types.

The Eastern Gangetic Plain (EGP) as known for high weed seed bank in soils particularly lower EGP (Bangladesh and West Bengal) because of blaze and consistent floods get weed seeds from distance their catchment regions.

The weed species likewise profoundly rely upon seasons as maize is developed round the year in three seasons for example Fall/autumn (Monsoon; June–September), Rabi (Winter; November–May) and Fall/ autumn 1 (Spring; February–June). The high weed seed pressing factor and variety make harder to oversee weeds physically in protection horticulture just as additionally in ordinary farming.

Grasses weed species during Fall/autumn season in the maize field in EGP are Echinochloa colona (L.), Digitaria ciliaris (L.), Leptochloa chinensis (L.), Dactyloctenium aegyptium, Cynodon dactylon, Echinochloa crus-galli, Eleusine indica, Setaria viridis, Panicum javanicum, Paspalum commersonii etc.

The broadleaf weeds are Marsilea minuta, Polygonum hydropiper, Galinsoga ciliata, Physalis heterophylla, Heliotropium indicum, Phyllanthus niruri, Euphorbia hirta, Jussiaea repens, Amaranthus spinosus, Amaranthus viridis, Spilanthes paniculata, Lindernia anagallis, Paspalum distichum etc. and the major sedges weeds are Cyperus rotundus, Cyperus difformis, Cyperus iria, Eclipta prostrata, Ludwigia octovalvis, Portulaca oleracea, Fimbristylis miliacea, Scirpus spp. etc.

The perennial weeds species dominant round the years but the seasonal weeds are different when maize grown in dry season (winter maize) and the major dominant weeds are Polygonum persicaria, Polygonum pensylvanicum, Polygonum orientale, Oldenlandia diffusa, Oldenlandia aquatic, Oxalis corniculate, Chenopodium arvensis, Physalis minima, Solanum nigrum, Hydrocotyle ranunculoides, Ageratum conyzoides, Medicago denticulate, Avena ludoviciana etc.

A field experiment found that the most dominant weeds in maize field were sedge C. rotundus and dicot weeds T. portulacastrum, D. arvensis, P. niruri and grass C. dactylon. In USA, it observed the most common broadleaves weed species in a maize crop were Chenopodium album, Asclepias syriaca, Ambrosia artemisiifolia, Physalis heterophylla and Polygonum rensylvanicum and the most common grasses were Elymus repens and Setaria pumila.

Figure 6.4. Major weed species in association with maize

Source: Image by Pxhere

In another studies conducted on sandy loam soils at Hyderabad, India, revealed that among the major weeds in maize field E. colona (grasses), E. crus-galli, Paspalum distichum, C. rotundus (sedges), Ageratum conyzoides and T. portulacastrum, Sonchus oleraceus, Acalypha indica, Eclipta alba and Parthenium hysterophorus (broad-leaved) were the predominant weeds.

A 3 years field survey in Andhra Pradesh of India also confirmed that most dominant weed species in maize field were E. colona, followed by P. repens, T. portulacastrum, and D. arvensis.

6.4. THE CRITICAL PERIOD OF CROP-WEED COMPETITION

The basic harvest development stages considers as the weakest period for crop-weed contest, during which harvest should be weed free to forestall yield misfortunes. Prior examinations noticed that the basic time of weed control in maize goes from 7 to 56 days in the wake of seedling development. Different investigations likewise announced, the basic period ordinarily relates for maize up to 8–10 leaf stages.

More extensive covering dispersing and moderate developing nature of the maize harvest should control weeds in first till 21–28 days subsequent to planting for nothing from crop-weed rivalry and it was additionally recommended that if the weeds are not control inside the basic harvest development organizes, the yield misfortunes may happen 30–100%.

Weed species, densities, and their collaborations impact maize yield misfortune. Weed plants (Figure 6.1) contend with maize for their fundamental development assets like water, supplements, space and so forth which at last lessen the yield up to 65% when weeds control measure was not performed at basic harvest development stages.

While, some dangerous weeds species as they are comparative in nature and life pattern of maize are hard to control. A well-known researcher named as Massinga announced that the yield decrease in maize could be 91% by rivalry if in excess of eight amaranth (Amaranthus palmeri S. Wats) plants per meter line length.

Figure 6.5. An illustration of cannabis plant

Source: Image by Wikimedia commons

6.4.1. Weed Control in Zero-Till Maize by Chemical Measures

In maize creation, weed management is considered as a significant agronomic measure for achieving the expected yield. To limit the maize yield misfortune because of weed rivalry, ranchers are rehearsing a few strategies for controlling the weeds are accessible like mechanical, social, natural and synthetic control techniques.

The social techniques are over the top expensive and tedious in this way, ranchers need to move towards other elective strategies for weed control. Besides, because of the expanding cost and non-accessibility of work for manual weeding during top and basic maize development organizes fundamentally impact the maize yield.

The job of herbicides isn't just controlling the weeds convenient and adequately, yet additionally offers an incredible breadth for limiting the expense of creation. The synthetic control strategy is speedy, more successful, time and work saving technique than others.

Notwithstanding, it is critical to utilize an expansive range herbicide program including pre-and post-development herbicides for season-long successful weed control and to stay away from shifts towards tricky weed species or advancement of herbicide-safe weed biotypes.

Then again, it is definitive to choose the suitable weedicide relying on the weed greenery exist in a given field. Moreover, the exact portion, strategies, weeds development stage, timings, soil dampness and application methods ought to be followed.

Figure 6.6. Weed control in zero-till maize by chemical measures

Source: Image by Pixabay

Various herbicides have been assessed in successive mix and recommended pre-and post-development herbicide application for powerful weed control in dry direct drill-cultivated rice frameworks, including under zero-culturing conditions.

Utilization of pre-planting, pre-and post-development use of herbicides would make herbicidal weed control more satisfactory to ranchers which won't change the current agronomic practices however will consider full oversight of weeds under CA based administration rehearses.

Pre-rise herbicides shower will control the weeds as long as 25 days in the wake of cultivating and followed that post-rise application relying upon weed vegetation will take care to keep weeds beneath crop injury level. Pre-planting, pre- and post-rise herbicides either in successive or tank combination will be dealt with a wide range of weed verdure an optimal methods taking into account financial aspects and helpfulness in maize.

6.5. WEED MANAGEMENT PROGRAM IN MAIZE

6.5.1. Social Control

Social practices assume a significant part in weed the board program in corn. Corn is an extremely aggressive harvest; so whenever oversaw appropriately, it gives significant contest against weeds. Exploration has shown that weeds that arise following a month of corn foundation lessly affect corn yield; consequently, early-season weed control is critical to get a serious corn yield.

It is imperative to set up a uniform plant remain at wanted thickness. Soil tilth, fruitfulness, pH, and seepage should be appropriate for the yield to be cutthroat with weeds. However much as could reasonably be expected, the yield should be figured out how to limit weights on the harvest from creepy crawly and infection harms and ecological burdens (ice, flooding, dry spell, and so forth) Line separating is a significant social work on influencing weed control since corn in thin lines will conceal soil surface sooner than corn in more extensive columns.

When the shelter has shut, next to no light arrives at the dirt surface or weeds underneath the overhang. The worth of early overhang conclusion for weed control is particularly obvious when weed control program in corn is subject to post emergence herbicides as it were. Truly, crop revolution has been perhaps the most well-known strategies for overseeing weeds.

The more assorted the harvest in pivot in planting time, development propensity, and life cycle, the more compelling the turn will be in controlling weeds. In this way, the choice of a harvest in pivot that incorporates little grains, scavenges, and vegetables is critical; in any case, such yields are not, at this point generally filled in the North Central USA.

Utilization of cover crops is another illustration of social control of weeds. Cover crops can be used for a variety of purposes including protecting the soil against erosion, improving soil structure, fixing nitrogen, feeding the soil biological life, and managing soil moisture.

A key soil wellbeing idea is that there ought to be something green and developing during however much of the year as could reasonably be expected. Grasses give the dependable buildup cover since they have a higher carbon to nitrogen proportion in their biomass contrasted with non-grass species. Furthermore, they improve snow get in the colder time of year and decrease wind disintegration in the spring contrasted with the exposed soil.

Taller brassicas with broad leaves like rape, mustards, and canola will also effectively reduce wind erosion and catch snowfall, but they provide less residue. In conclusion, a healthy, vigorous corn crop with a high yield potential will be very competitive with weeds; however, competition from the crop alone is not sufficient to provide a season-long weed control. Other methods of control must be used in conjunction with cultural control measures.

6.5.2. Mechanical Weed Control

Tillage is the most common method of mechanical weed control and it can be divided into two categories: (1) preplanting tillage and (2) in-row cultivation. The reason for preplanting culturing is to kill every one of the weeds present prior to planting corn to give the yield a superior begins to rival weeds during the underlying stage. Field cultivators and circles are usually utilized by producers, and they are profoundly successful for controlling weed seedlings whenever utilized appropriately.

The in-line development is utilized to eliminate weeds after the harvest has been planted, generally utilizing turning scraper or an interrow cultivator. Rotary hoes are most effective on small-seeded broad-leaved weeds and grasses, but they are less effective on large-seeded broad-leaved weeds, such as giant ragweed, velvetleaf (Abutilon theophrasti Medik.), cocklebur (Xanthium strumarium L.), etc.

Rotating hoes are generally worked at the speed of 13–19 km/h and ought to be utilized subsequent to planting the harvest however before weeds have arisen or after weed germination. Another benefit of in-line development is that they are valuable when soil-applied herbicides neglect to control weeds because of absence of precipitation.

A few sorts of in-line cultivators are accessible on the lookout, however it is imperative to change the gear to adequately kill whatever number weeds as could be allowed in the interrow region while limiting the unsettling influence of the yield plants.

6.5.3. Flame Weeding

Blazing controls weeds basically by cracking the phone films that prompts ensuing tissue parching. Propane burners can create ignition temperatures up to 1900°C, which raises the temperature of the uncovered plant tissues quickly.

An expansion of temperature above 50°C inside the plant cells can bring about the coagulation (denaturation and collection) of layer proteins prompting loss of the film trustworthiness. Therefore, blazed weeds would pass on or their serious capacity against the harvest would be seriously diminished. The vulnerability of plants to fire to a great extent relies upon their warmth evasion, heat resilience, or both.

The degree to which heat from the flares enters plants relies upon the blazing method and leaf surface dampness. The impacts of flaring on plants are affected by a few components including temperature, openness time, and energy input.

Depending on the exposure time, protein denaturation may start at 45°C. Temperatures in the range of 95–100°C at least for 0.1 s have been reported to be lethal for leaves and stems. Heat from the flames has a direct effect on the cell membranes and an indirect effect on the subsequent tissue desiccation.

Cellular death after fire treatment is fundamentally because of initial thermal disruption of cellular membranes rapidly followed by dehydration of the affected tissue. Tissue drying out happens chiefly because of extension of the cell substance (made of up to 95% water), ensuing blasting of the cell films, and coagulation of layer proteins.

The viability of fire weeding was accounted for to be affected by a few variables, including the presence of defensive layers of hair or wax and lignification, the actual area of the developing point at the hour of flaring, plant development stages, the regrowth capability of plant species, the strategy of blazing, and the overall leaf water substance of plant species.

Ulloa et al. conducted a series of studies where the authors intentionally flamed several agronomic crops such as field corn, popcorn (Z. mays L. var. everta), and sweet corn (Z. mays L. var. rugosa).

Reaction to communicate blazing changed among corn types, their development stages, and propane portion. Popcorn was the most un-open minded while field corn was the most lenient to communicate blazing dependent on the greatest yield decrease acquired with the most elevated portion of propane (85 kg ha−1).

Field corn flared broadcast at the five-leaf stage (V5) was the most lenient while the two-leaf stage (V2) was the most powerless, which had the most noteworthy visual harvest injury and the biggest loss of yield and yield segments.

Visual yield injury indications included introductory brightening and afterward cooking of leaves. Stunting of growth was especially evident when the plants were flamed with higher propane doses (44 and 85 kg ha−1). Most visual harvest wounds, nonetheless, were transient as corn plants had all the earmarks of being outwardly recuperated inside half a month.

Popcorn flared at the V5 stage was the most lenient while the V2 was the most vulnerable stage for broadcast blazing. Plants blazed at the V2 stage had the best return misfortune and the most reduced yield segments.

This may be clarified by the way that the ear and tuft tissues are not separated at the V2 stage; consequently, presenting the plants to the pressure from warmth can bring about conceivably more limited cobs. In examination, flaring popcorn plants at later development stages (e.g., V5 or V7) affected cob size as the ear and decoration tissues begin to separate at the V5 stage, and by the V7 stage, cob and tuft sizes are as of now foreordained.

A propane portion of 60 kg ha−1 brought about 8, 9, and 21% yield decreases at the V5, V7, and V2 stages, individually, which would not be adequate by natural ranchers. These yield decreases were the aftereffect of the deliberate flaring where lights were situated straight over the harvest lines.

Be that as it may, situating blazes beneath the popcorn shelter would lessen the openness time to the warmth and, subsequently, ought to diminish popcorn yield misfortunes. Sweet corn blazed at the V7 stage was the most open minded while the V2 was the most un-lenient stage for broadcast flaring.

Sweet corn blazed at the V7 stage had the least yield misfortune and the most un-influenced yield parts contrasted with plants flared at the V5 and V2 stages. The V2 was the most delicate stage for broadcast blazing, bringing about the best return misfortune and the biggest consequences for yield parts.

Among the yield parts, number of plants per square meter and seeds per cob were the most influenced boundaries when blazing was led at the V2 and V5 stages. Sweet corn for the most part begins to speed up its development around the V6–V7 stages (developing point arrives at soil surface).

This development speed increase in sweet corn is likewise combined with expanding convergence of sugars in cell and stem tissues, which requires more energy to bubble water in the cell. A propane dose of 60 kg ha−1 caused yield losses of 6, 11, and 20% for the V7, V5, and V2 stages, respectively. From a practical standpoint, the 6% yield reduction of sweet corn flamed broadcast at the V7 stage may not be acceptable by organic growers.

Nonetheless, yield decreases were the consequence of the deliberate flaring straight over the harvest. An option may be to coordinate the fire beneath the harvest shelter to save foliage from the warmth, which could bring about lower yield misfortunes (e.g.,< 5%).

Comprehend that propane flaring ought not be the lone technique for nonchemical weed control; it's anything but a piece of an IWM program. Different measures are as yet expected to control weeds that arise later during the developing season.

More exploration is expected to maybe foster new flaring hardware and strategies, or to inspect distinctive situating of the burners to keep away from any huge harvest harm and yield decreases. Data from such examination would grow flaring choices as a component of an IWM program for both natural and traditional harvest production frameworks.

6.5.4. Natural Control

The organic control approach utilizes the weed's normally happening adversaries to help lessen the weed's effect on agribusiness and the climate. It just intends to rejoin weeds with their regular adversaries and accomplish maintainable weed control.

These regular adversaries of weeds are frequently alluded to as organic control specialists. For instance, a business bio-herbicide Colego, a contagious herbicide, has been utilized to control northern joint vetch (Aeschynomene History of the U.S L.) in rice (Oryza sativa L.) in the southern USA.

It is important that the natural control specialists don't become insects themselves. Extensive host-particularity testing is required according to numerous administration rules and guidelines preceding the arrival of natural control specialists to guarantee that they won't represent a danger to nontarget animal categories, like local and rural plants.

Not all weeds are appropriate for natural control. Fostering a natural control project requires a considerable venture, here and there costing a large number of dollars. Right now, there are no business items for natural weed control in corn, however this region offers incredible potential for new weed control alternatives later on.

6.5.5. Chemical Weed Control

Use of herbicides is the main technique for weed control in corn. Herbicides have been received by a greater part of corn cultivators in the USA and numerous different pieces of the world since they are powerful and affordable.

Herbicides can be applied at different time intervals, such as before the crop is planted (preplant), after the crop is planted but before emergence (pre emergence), and after crop emergence (post emergence). The choice of herbicide application timing depends on many factors and varies from grower to grower and field to field. Many corn growers use more than one herbicide applications that may provide a season-long weed control.

6.6. EFFECT OF PRE-SOWING HERBICIDES ON WEED CONTROL OF MAIZE

Generally, pre-sowing/planting herbicides are non-selective which are applied to control prevailing complex annual and perennial weeds flora erstwhile to planting, particularly under the CA-based cropping system. Among them, herbicide Glyphosate (1 kg ai ha^{-1} or 0.5–1.5% by volume), Glucofosinate, or dicamba and Paraquat (0.5 kg ai ha^{-1} or 0.5% by volume) are the most widely applicable herbicides.

Glyphosate and Paraquat herbicides are accessible and well known as these are foundational non-particular and contact herbicides, separately, and kill both yearly and perpetual weeds. To be successful, the foundational

(Glyphosate) ought to be applied when weeds are developing effectively with the goal that the herbicide is retained and moved into the whole plant framework, however contact herbicide (Paraquat) can be applied not long before the planting.

Right now, the current paces of numerous herbicides don't work as expected against numerous weeds, because of their opposition against the objective weeds. Along these lines, it is critical to apply another item to be accessible sooner than ordinary to amplify the commitment of leftover weed control after crop development.

The previous investigation found that if the pre-planting herbicide is applied prior to planting and is joined in the dirt with light culturing, the viability of the applied herbicides was tracked down the greatest.

6.7. IMPACT OF PRE-RISE HERBICIDES ON WEED CONTROL IN MAIZE

At the point when herbicides apply quickly or 1–4 days after maize seeds planting, and it very well may be likewise blended in with soil during planting, yet it should be applied before weed seed development, are known as pre-rise herbicides.

Prior discoveries showed that pendimethalin can be applied as pre-development to gain greatest weed power proficiency and harvest selectivity by diminishing the weed populace and expanded the maize grain yield over the weedy check field.

According to a well-known researcher named as, Mekky revealed that when pre-rise herbicides were applied following seed planting or pre-rise, weed control productivity was the greatest and furthermore expanded the maize yield.

A field explore in the mud topsoil soils of Guntur, Andhra Pradesh (India), with zero culturing maize found that atrazine 1.5 kg ha−1 applied at pre-rise followed by (fb) manual hand weeding (HW) at 30 DAS recorded the tallest plant and the most extreme dry-weight over un-weeded check at all phases of harvest development.

Researcher detailed that when maize was become under zero till condition with pre-development herbicides atrazine at 1.0 kg ha−1 and topramezone 0.030 kg ha−1 in blend of two hand weeding at 20 and 40 DAS was delivered fundamentally higher plant tallness, dry matter production of maize.

Figure 6.7. Impact of pre-rise herbicides on weed control in maize

Source: Image by Pix4free

6.8. EFFECT OF POST-EMERGENCE HERBICIDES ON WEED CONTROL IN MAIZE

Application of herbicides after the emergence of maize and weed are well-known as post-emergence herbicides. Generally, post-emergence herbicides spray/apply in standing crop targeting weeds canopy by using the sprayer equipment.

Nonetheless, when a solitary herbicide is utilized for quite a while for controlling similar weeds it might create opposition against the particular weed(s). Consequently, long haul premise nonstop utilization of a similar herbicide ought to be kept away from. In this way, it ought to be pivoted with the utilization of numerous herbicides with an alternate method of activity to stay away from/postpone weed opposition against explicit weeds.

Researchers found that utilization of at least two viable herbicides use as a tank combination are more viable to control widen the range of weeds including grasses, broadleaves, and sedges than a solitary application.

In the USA, Atrazine has been utilized since 1958 out of a few million hectares of maize field because of its minimal expense, control of an expansive range of broadleaf weeds, adaptable application timing, and furthermore can be utilized as both pre-and post-rise in mix with a few different herbicides.

Because of long haul and consistent use, the atrazine was found in amassing in food items, groundwater, and sea-going frameworks. Along these lines, we ought to be mindful so as to know the leftover impact of applied herbicides when utilizing the as present moment just as long haul.

6.9. INTEGRATED WEED MANAGEMENT IN CORN

IWM is the control of weeds utilizing extraordinary, free techniques inside a framework instead of depending on a solitary strategy. The fundamental point of IWM is to lessen the determination pressure for the advancement of protection from any single strategy for weed control (Chauhan et al., 2017).

Sadly, weed research in many nations is arranged toward herbicide research. Successful weed management and a decreased danger of the development of herbicide-safe weeds relies upon additional examination into IWM across worldwide settings.

Integrated weed management (IWM) has been characterized as a multidisciplinary way to deal with weed control, using the utilization of various elective control measures. The IWM includes a blend of social, mechanical, natural, hereditary, and synthetic strategies for a viable and efficient weed control that diminishes weed impedance with the harvest while keeping up satisfactory yield yields.

None of the individual control measures can give total weed control. In any case, if different parts of IWM are carried out in an efficient way, critical advances in weed control innovation can be accomplished. The IWM approach advocates the utilization of all accessible weed control choices that include:

- Selection of a well-adapted crop variety or hybrid with good early-season vigor and appropriate disease and pest resistance
- Appropriate planting patterns/spacing and optimal plant density, improved timing, placement, and amount of nutrient application
- Appropriate crop rotation, tillage practices, and cover crops
- Suitable choice of mechanical, biological, and chemical weed control methods
- Alternative weed control tools (flaming, steaming, infrared radiation, sand blasting, etc.)

6.10. CONCLUSION

Weeds are one of the significant impediments in supporting the yield efficiency. Weeds contend with crops for supplements, soil dampness, sun-based radiation and space; and diminish the yield and nature of produce.

Additionally, they likewise go about as substitute host for bug irritations and infection causing creatures. Weed issues shift in various yields, seasons, agro-biological conditions and the executives rehearses. Various techniques for weed the board in field crops including preventive, social, mechanical, compound organic and biotechnological are being utilized with changing level of accomplishment.

Because of ceaseless utilization of single technique for weed control (particularly herbicide), weeds foster obstruction and become hard to control. Constant utilization of a similar herbicide for a more drawn-out period in a similar harvest in a similar region additionally prompts shift in weed verdure.

Thinking about the variety of weed issue and agro-biological systems, no single technique, regardless of whether manual, mechanical or synthetic gives the ideal degree of productivity under all circumstances. Different weed the board draws near, for example, the utilization of old seedbed rehearses, the revolution of culturing frameworks, and the utilization of harvest buildup as mulch, weed serious cultivars with high return potential, proper agronomic practices and fitting herbicide timing, turn, and mixes, should be coordinated to accomplish viable, maintainable, and long haul weed control.

Acquaintance of transgenic crops safe with non-specific herbicide like glyphosate gives ranchers the adaptability to control an expansive range of weeds with insignificant harvest harm.

Weeds are a significant biotic limitation to production in various editing frameworks. A solitary strategy for control won't give satisfactory long haul weed the board, rather frequently bringing about the advancement of obstruction.

Weeds are the reason for critical yield misfortune, even after the use of a specific control technique. There is a developing need to diminish this yield misfortune to take care of an always expanding human populace. In this way, there is a need to create successful and feasible IWM programs.

REFERENCES

1. Chauhan, B., 2020. Grand Challenges in Weed Management. Frontiers in Agronomy, [online] 1. Available at: <https://www.frontiersin.org/articles/10.3389/fagro.2019.00003/full> [Accessed 29 June 2021].
2. Hossain, A., Tanjina Islam, M., Shohidul Islam, M., Nurislam, Ahmed, S., Kumer Sarker, K. and Kumar Gathala, M., 2020. Chemical Weed Management in Maize (Zea mays L.) under Conservation Agricultural Systems: An Outlook of the Eastern Gangetic Plains in South-Asia. Maize - Production and Use, [online] Available at: <https://www.intechopen.com/books/maize-production-and-use/chemical-weed-management-in-maize-em-zea-mays-em-l-under-conservation-agricultural-systems-an-outloo> [Accessed 29 June 2021].
3. J. S. Mishra, A.N. Rao, V.P. Singh and Rakesh Kumar, 2016. Weed management in major field crops. [online] researchgate. Available at: <https://www.researchgate.net/publication/309771518_Weed_management_in_major_field_crops#:~:text=Various%20weed%20management%20approaches%2C%20such,combinations%2C%20need%20to%20be%20integrated> [Accessed 29 June 2021].
4. Jhala, A., Knezevic, S., Ganie, Z. and Singh, M., 2014. Integrated Weed Management in Maize. Recent Advances in Weed Management, [online] pp.177-196. Available at: <https://agronomy.unl.edu/documents/Integrated%20Weed%20Mana.%20in%20Corn.pdf> [Accessed 29 June 2021].
5. NIOS, 2013. Definition and Characteristics of Weeds - NIOS. [online] Oer.nios.ac.in. Available at: <http://oer.nios.ac.in/wiki/index.php/Definition_and_Characteristics_of_Weeds> [Accessed 29 June 2021].
6. Suzette Bezuidenhout, 2016. Weed management principles for maize. [online] Kzndard.gov.za. Available at: <https://www.kzndard.gov.za/images/Documents/RESOURCE_CENTRE/GUIDELINE_DOCUMENTS/Research_and_Technology_Bulletin/Weed-Management-Principles-for-Maize.pdf> [Accessed 29 June 2021].
7. TNAU, 2016. Weed Management :: Integrated Weed Management. [online] Agritech.tnau.ac.in. Available at: <https://agritech.tnau.ac.in/agriculture/agri_weedmgt_iwmintegratedweedmgt.html> [Accessed 29 June 2021].

Chapter 7

Agronomic Practices of Maize

CONTENTS

7.1. Introduction 146

7.2. Concepts And Guiding Principles of Good Agricultural Practices (GAP) 148

7.3. The GAP Approach and Concepts 150

7.4. Integrated Pest Management (IPM) 152

7.5. Integrated Crop Management (ICM) 153

7.6. Tips on Best Practice Maize Production 154

7.7. Maize Agronomic Principals 157

7.8. Conclusion 162

References 164

This chapter revolves around the agronomic practices of maize. It begins with the concepts and guiding principles of good agricultural practices (GAP), which involves sustainable agriculture. It then talks about the three major concerns, for the survival of a crop which includes the ecological concern that is the productivity of soil, economic and social concern which deals with the price of food and incomes of small and rural farmers, as well as the impacts on human health which is food safety and hygiene etc.

It further highlights the approaches and concepts of good agricultural practices then goes on defining the various approaches such as the risk assessment approach, preventive measures, continuous monitoring of the production process.

In the second half of the chapter, it talks about the integrated pest management (IPM) and the integrated crop management (ICM). It then discusses the different techniques which deals with the minimal tillage, mulching, high quality plant spacing, etc. Then it outlines the maize agronomic principles, the climate and location, response to nutrition, rotation and cultivation, planting, maturity and harvesting as well as crimping and mixing towards the end.

7.1. INTRODUCTION

Harvest yield potential can be expanded using suitable agronomic practices. Integrated agronomic practice (IAP) is a powerful method to build maize (Zea maysL.) grain yield and nitrogen use efficiency (NUE); be that as it may, the physiological cycles related with gains in yield potential acquired from IAP, especially the distinctive under different soil richness conditions, remain ineffectively comprehended.

An IAP system including ideal planting thickness, split compost application, and sub ruining culturing was assessed more than two developing seasons to decide if the impacts of IAP on maize yield and NUE vary under various degrees of soil richness.

Compared to farmers' practices (FP), IAP increased maize grain yield in 2013 and 2014 by 25% and 28%, respectively, in low soil fertility (LSF) fields and by 36% and 37%, respectively, in high soil fertility (HSF) fields.

The large yield gap was attributed mainly to greater dry matter (DM) and N accumulation with IAP than with FP owing to increased leaf area index (LAI) and DM accumulation rate, which were promoted by greater soil mineral N content (Nmin) and root length.

Post-silking DM and N accumulation were also greater with IAP than with FP under HSF conditions, accounting for 60% and 43%, respectively, of total biomass and N accumulation; however, no significant differences were found for post-silking DM and N accumulation between IAP and FP under LSF conditions. Thus, the increase in grain yield with IAP was greater under HSF than under LSF.

Food safety has acquired expanding significance throughout the years on account of its importance both from wellbeing and exchange viewpoints. The production of safe food is fundamental for shielding shoppers from the perils of food borne ailments and is significant both in the homegrown food business just as for expanding seriousness in trade markets.

Risks may happen at various phases of the natural pecking order beginning right from the essential creation, for example buildups above allowed levels, microbial foreign substances and hefty metals. It hence gets critical to address food handling directly from food production at ranch level.

Carrying out Good Agricultural Practices (GAP) during on-field production and after production measures bringing about safe rural items is of colossal significance for guaranteeing a protected food supply.

Many bringing in nations just as homegrown purchasers, particularly coordinated retailers, are expecting makers to carry out GAP as an essential for acquirement to guarantee the quality and wellbeing of their produce. What's more, carrying out GAP additionally advances manageable farming and adds to meeting public and worldwide natural and social formative goals.

It has been archived that execution of GAP energizes advancement of the ideal utilization of assets like pesticides, manures, and water, and eco-accommodating horticulture. Its social measurement is shielding the horticultural laborers' wellbeing from inappropriate utilization of synthetic substances and pesticides.

Large numbers of the SAARC nations had not embraced GAP and didn't have an institutional framework for supporting GAP as far as norms just as their execution, confirmation and accreditation. The majority of the food handling principles were centered on final results. It was thusly imperative to help the SAARC nations to foster GAP guidelines and to help them assemble a confirmation and accreditation instrument to empower the execution and check of GAP. With the above background, FAO, at the request of some SAARC countries, through a regional project on "Development of Standards and Scheme for Good Agricultural Practices (GAP) Implementation and

Certification in Countries of SAARC" developed a common standard on GAP for the horticulture sector along with a national system for its implementation.

This GAP scheme has been created as a regional plan that might be embraced all things considered or adjusted by nations to suit their necessities. The interaction for advancement of the plan was through starting conversations on a draft at a local workshop in September 2014 that was then carried out in four pilot nations, specifically Bangladesh, Bhutan, Maldives and Nepal.

The norm and public execution structure as far as a Scheme Owner (SO), certificate body (CB) and accreditation body (AB) framework was first settled upon and was trailed by help gave to the SO and CB to set up their cycles and frameworks dependent on the draft plot. In light of encounters in the four nations, the Scheme has now been amended. The Scheme can be utilized by any nation universally.

7.2. CONCEPTS AND GUIDING PRINCIPLES OF GOOD AGRICULTURAL PRACTICES (GAP)

7.2.1. Sustainable Agriculture

Figure 7.1. Sustainable agriculture

Source: Image by Flickr

Today, the partners of the agri-food inventory network incorporate insurance of the climate act as an indispensable piece of their arranging programs, practically identical to some other parts of their arrangements. They perceive that horticulture exercises are a huge factor in ecological debasement.

The utilization of the seriously restricted normal assets of good soil, new water, and cultivable land should be painstakingly delineated to stay away from inefficient use. The requirement for an always expanding pace of food production necessitates that we persistently work these valuable soils. This is the explanation that feasible horticulture is crucial.

Sustainable agriculture is a farming system that meets the need for safe, nutritious, and affordable food for the world population in a way that progressively conserves the environment and natural resources. It involves seeking to optimize skills and technologies to achieve long term productivity and profitability for stakeholders of the agricultural industry in order to ensure that future generations can also experience the same prosperity that we enjoy today.

From the definition above, three major concerns have emerged for farmers currently working on small plots of land who strive for the survival of their crops:

1. **Ecological Concerns**
   i) Soil productivity (erosion, depletion of top soil)
   ii) Water (depletion, groundwater usage, contamination)
   iii) Pest and disease resistance to pesticides
   iv) The greenhouse effect and climate change
2. **Economic and Social Concerns**
   i) The price of food
   ii) Incomes of the small and rural farmers
3. **Impacts on Human Health**
   i) Food safety and hygiene
   ii) The health and welfare of farm workers

How can economical agribusiness profit farmers when conditions for cultivating in such countless agricultural nations are in a particularly frantic state? Actually, if these ranchers sit idle, their lives won't ever improve. In the present market climate, these farmers may in any case have a little window that will permit them to partake in the standard of agri-food creation.

In spite of the fact that it is obviously a fact that these ranchers need support, regardless of whether it is specialized or monetary, what they need most are open doors on the lookout. Luckily, purchasers today are significantly keen on supporting ranchers who use maintainable methodologies.

Previously, the focal point of improvement financial matters fixated on little and provincial ranchers, who will in general be poor and nearly less useful. As of late, notwithstanding, there are numerous market analysts who contend that less consideration ought to be paid to little and country ranchers and that assets ought to be redirected to the bigger and more useful makers.

One new methodology that can be taken for these ranchers is the bunch framework for ranches under GAP, which hoists their hierarchical design and upgrades their production manageability.

This methodology has broad consequences for the work and social improvement openings for these networks, which surpass the advantages given by enormous scope, automated ranches. Moreover, little and provincial ranchers have certain benefits over bigger makers, for example, the capacity to be more mindful to their yield production and to deliver crops explicitly for specialty markets.

7.3. THE GAP APPROACH AND CONCEPTS

Food administrative specialists, retailers, and shoppers all have dynamic jobs in pushing food handling and practical food creation. Generally, it was the administrative specialists who assumed the most dynamic part by implementing corrective measures for guilty parties for resistance.

In any case, in the contemporary reality where purchasers are more educated and ready to set their expectations more vocal, they have now gotten more persuasive in directing more noteworthy food handling and practical production drives.

The part of checking and controlling principles and quality has now generally been supplanted by affirmations that reflect shopper requests and prerequisites from retailers. Hole is a fitting model to help ranchers address this adjustment of the advanced food production and circulation framework.

It depends on affirmations, peril investigation, and hazard appraisal. Hole moves toward the issues of protected and economical food production from the accompanying three levels:

7.3.1. Starting with a Risk Assessment Approach

Identifying hazards is a starting point for knowing where the dangers and risks lie for food safety and environmental damage. It is important to perform due diligence in order to know and understand the problems beforehand.

7.3.2. Preventive Measures

The removal of hazards (biological, physical, or chemical contaminants) from the Control Points (food operations procedures), and the prohibition of bad practices that potentially harm the environment (or farm ecology) will reduce or prevent accidents.

7.3.3. Continuous Monitoring of the Production Process

Consistent observation of the production interaction is fundamental since food tainting can happen under practically any conditions. Additionally, the possible effects of awful ranch rehearse on ranch nature should be perceived and consistently estimated to forestall the event of irreversible harm. As a part of feasible creation, this observing will guarantee that makers can reliably harvest enduring financial returns for their work.

The adoption of GAP into farm activities puts into practice two central concepts of agriculture that tend to be denied by farmers, particularly small and rural farmers. Farm extension agents should teach these two ideas to farmers in their introductions to the adoption of GAP.

- Agribusiness is a Science Farmers should attempt to comprehend the dangers implied for the natural capacities and synthetic outcomes of production frameworks. In doing as such, ranchers will find out about vegetation cycles and subsequently have the option to settle on choices on their next strategy or take legitimate reactions to any obstruction in crop cycles.
- Horticulture is a Business farmers should recognize that in business, there are chances required, just as rivalry. Ranchers should know about the occasions going on external their homesteads that could affect the promoting of their yields. Farmer should attempt some type of linkage into the market.

7.4. INTEGRATED PEST MANAGEMENT (IPM)

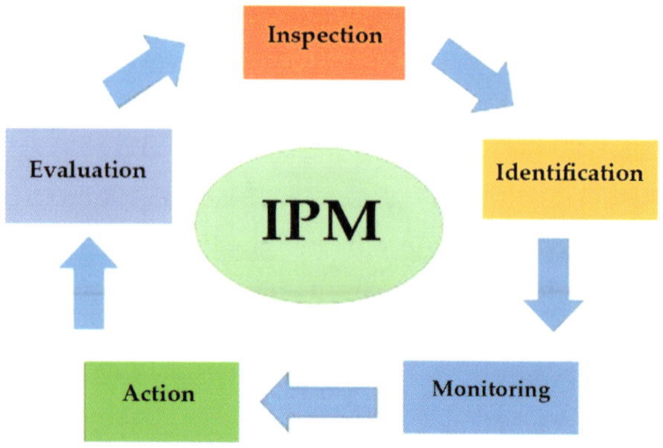

Figure 7.2. Integrated Pest Management (IPM) Cycle

Source: Image by IPM

Prior to the advancement of manufactured substance pesticides, ranchers rehearsed strategies to control vermin and sicknesses dependent on their perceptions of irritation and infection conduct and the impacts of the utilization of normally shaped synthetic mixtures. Ranchers additionally understood that the climate encompassing the homestead affected the control of irritations and illnesses.

The ranchers mentioned observable facts on plants that have bugs, bug repulsing plants, ranch sterilization, and the advantageous impacts of coordinated development methods.

The modern system of crop cultivation that emphasizes large-scale mono-cropping, high yield productivity and capital intensive and high-density planting has tended to take a different approach to pest and disease control.

A large number of the present ranchers consider nuisances to be the adversary and infection as an interruption into the production framework. They considered to be of synthetic substances as an end-all answer for annihilate nuisances and end plant infections.

The drawback of the imprudent utilization of substance pesticides with broad applications is that they bring about high buildup on the yield. Some substance pesticides keep up their intensity for extremely significant stretches. The diligent utilization of similar substance class of pesticides will bring about bother protection from the pesticides.

Farmers, who don't comprehend the science of pesticides, will keep on utilizing the synthetics, not understanding that they have gotten incapable. Some may even expand the volume of the pesticides in the mixed-up conviction that it will fortify their impact.

Such wrong applications cause a large group of issues for ranch biology. A portion of these impacts remember disturbing the biological equilibrium of nuisances for the homestead, the presentation of new irritations and infections, expanded bug protection from pesticides, and the annihilation of valuable insects like honey bees and hunters.

Integrated pest management (IPM) is a decision-making process that utilizes all available pest management strategies, including cultural, physical, biological, and chemical controls, to prevent economically damaging pest outbreaks and to reduce risks to human health and the environment.

The IPM approach doesn't preclude the utilization of manufactured pesticides. Notwithstanding, it specifies that preceding utilizing engineered pesticides, other non-intrusive techniques for bother the board ought to be made in a stride by step way.

Harvest exploring is one of the significant parts of an IPM program, fully intent on giving precise bug and yield improvement data on key insects and what they mean for the harvest. This remembers data for the occasional changes of the environment and the climate as these progressions affect irritations' life cycles, taking care of propensities, and conceptive practices.

Bug anticipation is the other part of the IPM program, which can be accomplished through a mix of social, physical, and natural controls to keep a bug invasion from arising and causing financial harm for the rancher.

Homestead tidiness practices with the expulsion and removal of plant squander forestall the production of fruitful justification for creepy crawly nuisances and plant sicknesses from germination. Pesticide control strategies ought to possibly be utilized when the above controls neglect to have an impact.

7.5. INTEGRATED CROP MANAGEMENT (ICM)

The objective of ICM is to manage the crop production to yield a high quality, profitable crop in ways that maintain and enhance the ecological environment of the whole farm for the wildlife and the community in and surrounding the farm.

The capacity to create an effective yield requires the total association of plant sustenance, soil the executives, crop assurance, tree the board, crop revolutions, ranch the board, and homestead disinfection.

To support the production crop cycles to yield a similar quality at a similar operational productivity, the hierarchical administration, waste and contamination the board, and untamed life and seen the executives are additionally impressive commitments to the production interaction.

Farmers are encouraged not to utilize mono-trimming in their homesteads. Under the ICM way to deal with development, ranchers are urged to incorporate as numerous free harvests/plants as could be expected, making biodiversity in their fields. In the image on the right, a farmer is developing mangos, however he has likewise joined lemongrass crops in the intercropping.

Lemongrass critically affects the bug bothers that assault the mango crop. The GAP prerequisites for crop the executives' tasks require hazard appraisals to be depicted for these homestead exercises affecting the climate of the ranch and the regions encompassing the homestead.

7.6. TIPS ON BEST PRACTICE MAIZE PRODUCTION

Figure 7.3. Best practices for maize production

Source: Image by FAOUN

The best strategy is known as Conservation Agriculture (CA), which is referred to in certain populaces as "Cultivating God's Way".

The highlights of this technique are:

- Minimal tillage. YOU SHOULD NEVER USE A PLOW, so no draft power is needed to prepare your fields. Instead of compacting or disrupting the soil's natural structure, you make precise planting holes in a consistent pattern throughout the field.
- Mulching to hold the dirt's dampness and to shield the dirt from disintegration. You ought to never consume vegetation covering the dirt. Essentially, try not to leave the dirt exposed, on the grounds that dampness will be lost and disintegration is bound to happen. All things being equal, brush, cut, or cut the plants and past stover, permitting it to spread all the more uniformly over the dirt.
- Precise, high-quality plant spacing. Plan out your field and measure where each planting hole will be for the best overall distribution.
- Precise distribution of fertilizer and seed. Broadcasting seed and fertilizer on to the soil can be wasteful, and often results in poor soil-seed contact, lost fertility and ultimately lower yields.
- Planting in early- to mid-November, or immediately after the first measurable rainfall of the applicable growing season to ensure even germination and make the most of the rainwater.
- Thinning out/removal of weaker germinated plants, maximizing yield.
- Maintain disciplined, regular weed control. Weeds steal the moisture that your crops need.

Coming up next is a step-by-step manual for non-mechanized Conservation Agriculture:

Stage 1: Early in winter months going before your growing season, plan the format and harvest revolution of your territory. Decide the measure of land that will be utilized, and compute how much compost and seed will be required. In the event that you intend to sell the maize upon gather, have an overall arrangement for where it will be showcased, how it will arrive, and for the amount you anticipate that it should sell.

Stage 2: Start setting up the property as arranged, during the dry season and a long time before the anticipated beginning of the downpours. Control weeds all through the interaction. A consistent work process throughout a little while will guarantee that you are prepared at the legitimate time. Cut the grass, stover, and weeds and untruth those equally over the dirt as mulch.

Decisively measure out the fitting separating with gauges; on the other hand, you may utilize two stakes joined by a string with hitches at the distance where each planting opening ought to be burrowed. Those openings ought to be a ways off of precisely 60 cm by 75 cm. Utilize the outline underneath as an aide.

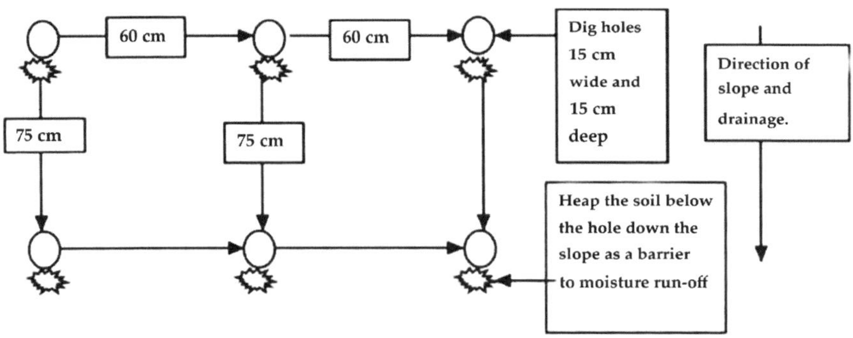

Figure 7.4. Openings for non-mechanized conservation agriculture

Stage 3: Apply basal compost necessities before the downpours start. A soda pop or brew bottle top can be utilized to apportion the measure of Compound D compost for each opening. Cover the manure with a modest quantity of soil to guarantee there is no contact with the seed, which would cause compound consume. In the event that you don't approach compost or can't manage the cost of it, you can utilize a full cup of very much treated the soil excrement in each opening.

Stage 4: Plant following the main quantifiable precipitation. Select an assortment of seed that suits the common precipitation got in your locale. In dry regions, utilize short-season assortments; possibly pick a long-season type on the off chance that you approach water system or you are in a space of extremely high precipitation. At the point when vital, every one of the openings can be watered by can, watering can or hose and left to settle. While the dirt is as yet sodden, place three seeds in each opening in a uniformly separated line, with one pip on each side and one in the center. Cover the seeds utilizing the dirt heaped close to the opening to make a level surface, with substantial soil lumps or stones covering the seeds.

Leave the mulch cover flawless between the openings. It is ideal to finish planting in one day, to support even germination and later, an even yield overhang, and which will assist conceal with excursion development of weeds.

Stage 5: The first round of weeding ought to be finished fourteen days subsequent to planting, and finished with at least soil unsettling influence. Digger weeds when they are little. Also, eliminate the most fragile seedling from each opening a little while after germination, which will leave a normal of two seedlings for every opening. In the event that just one seed sprouts in an opening, every one of the three seedlings can be left in the nearby opening.

Stage 6: After the maize plants are grounded and have developed to the three-leaf stage, ammonium nitrate manure ought to be applied. Your field ought to be weeded both when top-dressing, since you don't need weeds to go through and develop from the manure expected for your harvest.

A soda pop or lager bottle top can again be utilized to allot the AN, with one proportion of A for each planting opening. AN unquestionable requirement be applied to wet soil, so it is ideal to apply it not long previously or after it downpours. Apply the ways off of 5 - 10 cm from the plants, and don't allow the compost to contact the plants. On inclining ground, the AN ought to be applied to the dirt over the planting opening.

Stage 7: Harvesting among March and June will forestall grain misfortune because of birds. After the cobs have been eliminated, leave the excess stover lying in the field, which will ensure the dirt and improve its quality. Leave establishes in the dirt to separate and further improve the dirt's construction. Cut any weeds following harvest to forestall seed creation.

At last, plan openings similarly situated as the past season; these openings can be reused, with minor fixes, for your next developing season. Soybeans, sugar beans, and peanuts are suitable vegetables to turn with maize.

7.7. MAIZE AGRONOMIC PRINCIPALS

No other grain crop utilizes sunlight as effectively as maize and its yield per hectare is the highest of the grain crops.

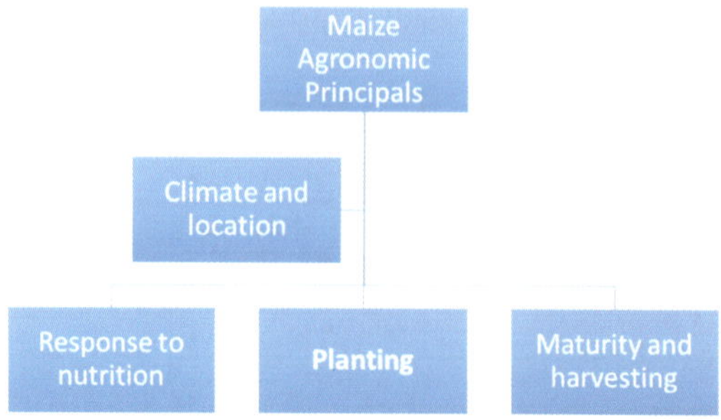

Figure 7.5. Maize agronomic principals

7.7.1. Climate and location

Figure 7.6. Worsen climate impact on maize production

Source: Image by MIT News

Maize flourishes best when air temperatures in the mid-year months are between 21-27°C so maize is basically developed between the scopes 55°N and 45°S.

Soils should be adequately warm to guarantee great solid rise with little check in development. The base temperature for germination is 10°C and at a dirt temperature of 16-18°C, maize regularly arises inside seven days of planting. Maize doesn't endure water logging, so great seepage is fundamental, particularly in mild areas and on heavier soils.

After development, each leaf stage is characterized by the highest leaf whose leaf collar is noticeable. Be that as it may, from around V6 onwards the lower leaves may bite the dust and tumble off, so at this stage it is critical to tally the quantity of collars on the stem.

7.7.2. Response to nutrition

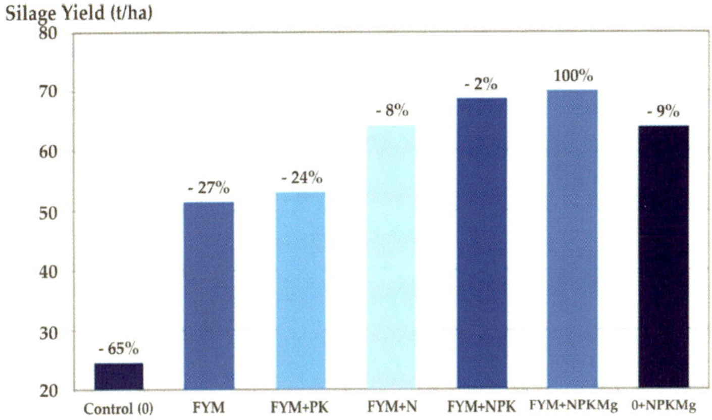

Figure 7.7. Effect of nutrition on forage maize yield

Source: Image by Yara – Germany – 2012

Table 7.1. Treatment details

	Nutrient (kg per ha)			
	N	**P_2O_5**	**K_2O**	**MgO**
Farm yard manure (FYM)	90	56	141	34
FYM + MPKMg	180	73	157	40
0 + MPKMg	170	70	150	35

Source: Table from Yara – Germany – 2012

At beginning phases of plant improvement, development is restricted by the size of the leaves and the root framework. Early leaf production additionally influences grain creation. Thus, all supplements should be promptly accessible right now or helpless development will confine yield. The basic development stage as for crop nourishment is the V6 stage.

7.7.3. Rotation and cultivation

While maize can be edited ceaselessly, pivot with different harvests assists with expanding yields by limiting infection, pest and weed occurrence. Past harvests, for example, vegetables can likewise decrease the requirement for nitrogen manure.

Animal manures and slurry provide valuable nutrients and at the same time helps to increase soil organic matter.

It is imperative to consider this remaining sustenance when arranging a manure program for the yield. Supplement shortages should be enhanced and preliminaries show huge reactions to applied compost over the utilization of fertilizer.

7.7.3. Planting

By and large, maize is planted 3 to 5cm somewhere down in columns, with separating administered by reaping gear, territorial inclinations and maize type/end use.

Grain maize is generally developed with column spacings of between 50 to 210cm. Closer line dispersing permits plants to abuse more dampness, supplements, and light prior because of more prominent space between plants in the column.

For grain maize grown in more arid regions, crop density can be as low as 25,000 plants/ha and in humid, tropical regions populations can be in excess of 75,000 plants/ha.

In temperate districts, maize for forage is usually planted at 90,000 to 110,000 seeds/ha and the best yields come from low height, sun-confronting fields.

It is normal practice to sow 10–15% more seed than is needed to consider decreased foundation and produce the ideal plant populace. Germination figures are generally around 95%, and another 5–10% of seeds might be lost to bugs, illness, or weeds.

7.7.4. Maturity and harvesting

Grain maize is reaped when grain has dried down to a fitting level. Whenever this has been accomplished yields can be left in the field and gathered sometime in the not-too-distant future. In some cases, particularly more temperate climates, harvested grain needs to be dried to 15% moisture or less, before it is stored in order to maintain grain condition and reduce mycotoxin build-up such as, aflatoxin.

Figure 7.5. Harvesting of maize

Source: Image by Mississippi crop situation

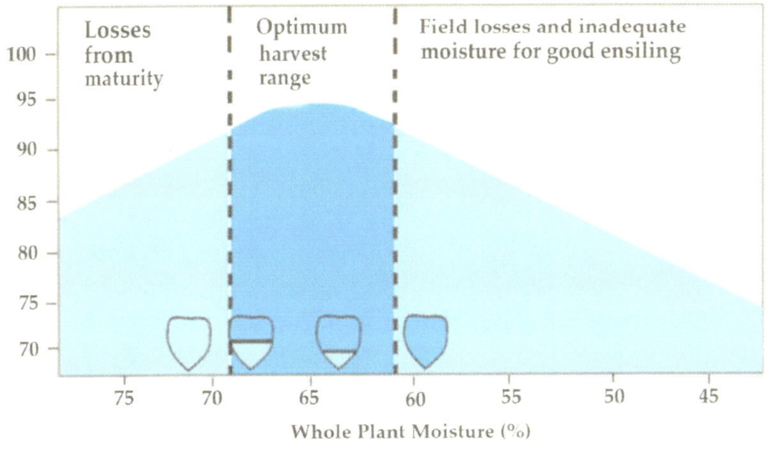

Figure 7.8. Optimum harvest stage

Source: Image by North Dakota State University – USA

Maize for silage is taken when the plant is as yet green and the grain is less experienced than that needed for grain maize. The best reap timing is at 28-35% DM or when the grain arrives at the half milk line stage – this gives the best trade off as far as silage amount and quality.

7.7.5. Crimping and mixing

Crimping, where the maize is left in the field until later and just the grain is harvested and run through a crimper before being ensiled, provides a high energy feed without the need for drying it down to levels of that required for storing grain maize.

Corn cob blending is an elective collect strategy utilizing consolidates with an adjusted header to reap corn and part of the cob. This blend is then ground and ensiled. It permits the maize to be taken before and gives an elective high energy feed.

7.8. CONCLUSION

The ever-increasing human populace along with practices like slash and burn, and the new expansion of surface mining has jeopardized moving development as rehearsed in the past by most ranchers, as methods for saving the dirt and looking after efficiency (Ekboir et al., 2002).

Because of the above reasons, ranchers keep on developing maize on a similar real estate parcel many more than one season, as well as rehearsing the slash and burn strategy that had added to the expanding land corruption. There is in this manner the need to give safe mediations to ranchers to create economically.

The utilization of no-till innovation monitors dampness by diminishing dissipation and overflow builds invasion of water and diminishes soil temperature changes. It additionally supplies supplements and natural matter from disintegrating mulch and improve soil faunal action (Goddard et al., 2008).

Boahen et al. (2007) expressed that the no-till idea is for resource saving horticultural production that endeavors to accomplish a satisfactory benefit along with high and supported production levels while simultaneously saving the climate.

Bonsu (2001) expressed that deposits left on the field during no-till fill in as mulch which shields the dirt from soil disintegration and breaking point weed development all through the developing season. Ranchers generally endeavor to supplant the ungerminated maize by re-cultivating at a similar spot or slope of dead plant(s).

In any case, these late developed plants endure various troublesome development conditions (Asiedu et al., 2001; Van Gassel et al., 1996).

Endeavors by certain people and establishments to improve plant populace thickness in No-till fields have been finished utilizing date of planting, topping off with seed maize, planting profundity, insect poison and fungicide treatment, firming planting spot and planting through mulch.

These were created to shield planting spot from the rodents, yet ranchers keep on experiencing loss of plant remains in their maize fields including through obliteration of seeds by rodents and seed borne microbes.

REFERENCES

1. Akowuah, P., Sarkodie-Addo, J. and Boa, K., 2012. Effect of Some Agronomic Practices to Increase Maize Yield in Ghana. Journal of Science and Technology (Ghana), [online] 32(2). Available at: <https://www.researchgate.net/publication/272327109_Effect_of_Some_Agronomic_Practices_to_Increase_Maize_Yield_in_Ghana> [Accessed 30 June 2021].
2. Chan, K., 2016. Manual on good agricultural practices (GAP). Japan: Asian Productivity Organization, Available at: <https://www.apo-tokyo.org/publications/wp-content/uploads/sites/5/Manual-on-Good-Agricultural-Practices-2016.pdf> [Accessed 30 June 2021].
3. FAO, 2016. A scheme and training manual on good agricultural practices (GAP) for fruits and vegetables. [ebook] Bangkok: FAO. Available at: <http://www.fao.org/3/i6677e/i6677e.pdf> [Accessed 30 June 2021].
4. Precision Agriculture, 2018. Maize agronomic principals | Yara UK. [online] Yara United Kingdom. Available at: <https://www.yara.co.uk/crop-nutrition/forage-maize/maize-agronomic-principals/> [Accessed 30 June 2021].
5. Zhou, B., Sun, X., Wang, D., Ding, Z., Li, C., Ma, W. and Zhao, M., 2019. Integrated agronomic practice increases maize grain yield and nitrogen use efficiency under various soil fertility conditions. The Crop Journal, [online] 7(4), pp.527-538. Available at: <https://www.sciencedirect.com/science/article/pii/S2214514119300248> [Accessed 30 June 2021].

Chapter 8

Integrated Pest Management in Maize

CONTENTS

8.1. Introduction 166

8.2. Diseases Pest Management Approach 168

8.3. Management Practices for Insect Pests 170

8.4. Pest Management 172

8.5. General Guidelines for Management of Resistance 180

8.6. Safety Parameters in Pesticides Usage 181

8.7. Insect Pest Management In Maize 182

8.8. Conclusion 189

References 190

This chapter sheds light on the integrated pest management in maize, which includes the management practices for insect pests, such as the cultural practices, genetic management, mechanical practices, bio pesticides, and biological control.

It then provides the highlights of identification and nature of damage, along with the features of identification and nature of damage. It also mentions the pre- planting practices, flowering to harvest, the natural enemies, and management strategies.

It also gives the general guidelines for management of resistance such as the insecticides and fungicides along with the *helicoverpa armigera* which is also known as corn earworm. Towards the end, it describes the safety parameters in pesticide usage and the insect pest management in maize.

8.1. INTRODUCTION

Pests are major biotic imperatives to accomplish independence in guaranteeing food security. Misfortunes because of pests fluctuate range 10-30% relying on the hereditary constituent of yield, its wellbeing and the administering climate. General public gauge of yearly harvest misfortunes because of pest adds up to ' 260000 million every year.

Be that as it may, carelessness of endemic regions can bring about complete harvest disappointments. Considering inefficacy of synthetic pesticides and natural issues thereof, Integrated Pest Management (IPM) has been acknowledged as a cardinal rule of Plant Protection in the general Crop Protection Program under the National Agricultural Policy of the Govt. of India. IPM being an eco-accommodating methodology, socially satisfactory and financially reasonable has been generally acknowledged the nation over.

The IPM package encompasses various management strategies for pest and disease problems. Pest monitoring is also one of the important components of IPM to take proper decision to manage any pest problem. It can be done through Agro-Ecosystem Analysis (AESA), field scouting, light, pheromone, sticky/yellow pan traps.

The Economic Threshold Level (ETL) of significant insects and sicknesses are likewise given in the bundle to enact suitable control measures on standing yields. The current bundle and practices was grown path back in 2001-02 by DPPQ and S, Faridabad providing food the need of augmentation personals in stretching out IPM strategies to ranchers.

Despite the fact that these were valuable, there is a need to refresh them considering changing environment and its effect on bothers and their security measures. A National Workshop on IPM for harmonization of Package of Practices was coordinated at the National Center for Integrated Pest Management, New Delhi, during 25-26th Feb., 2013 so as to give specialized information to the expansion functionaries and ranchers in the States.

The IPM package has been developed with the technical inputs from the experts from the PI (AICRIP), Indian Council of Agricultural Research (NCIPM), State Agricultural Universities, and DPPQ & S, Faridabad. It will likewise be valuable in decreasing the pesticide buildups in exportable rural items and would likewise help in the administration of bugs/illnesses/weeds/nematodes, which may get coincidentally presented in the country.

These bundles will be valuable for the analysts, augmentation laborers and ranchers the same who are occupied with the rural practices.

Maize (Zea mays L.) involves a significant spot in world farming. It is assessed that continuously 2020, interest for maize in agricultural nations will outperform interest for both wheat and rice. Maize, in India, positions fifth in complete region and third in absolute production and usefulness.

This degree of production must be considerably raised to fulfill developing need of maize for human food, creature and poultry feed, just as modern handling by the wet and dry mill operators to deliver value added items Maize can be successfully grown in rainy (fall/autumn), winter (rabi) and summer spring (zaid) crop seasons. Because of its divergent types, it is grown over a wide range of climatic conditions, ranging from near sea-level to several thousand meters above sea-level (2,700 m msl).

Maize can be filled in tropical, subtropical and mild environments, anyway tropical and sub-tropical involve a higher level of corn creation. It can likewise be filled in a wide range of soils going from sandy to substantial dirt.

Profound substantial soils are viewed as more appropriate taking into account their better water holding limit. Saline and soluble soils ought to be kept away from since these antagonistically influence crop development and advancement.

The prerequisite of manures relies on the situation with the dirt, past trimming history and length of the assortment to be developed. However, a balanced application of 60-120 kg N, 40-60 kg P and 40 kg K is recommended for various ecosystems. A population of 65,000-70,000 plants/ha is optimum for realizing high maximum yield. Among the factors adversely affecting productivity, ubiquitous prevalence of diseases and insect pests in the pre harvest stage are prominent.

The all-out financial loss of the harvest in India because of insect pests and diseases has been assessed to be of the request for 13.2%. Since there is for all intents and purposes no chance of expanding maize region, the efficiency must be raised by giving seed of improved cultivars, better agronomic practices and assurance against illnesses and bugs.

Subsequently, this report planned to give complete coordinated bug the board practices to diminish crop misfortunes brought about by infections and insects pests of maize.

8.2. DISEASES PEST MANAGEMENT APPROACH

There are more than (67) meanings of IPM, gave by governments, research associations, NGOs, and colleges (Bajwa and Kogan, 2002). Some expect that IPM will dispense with the utilization of yield assurance items uncommonly the substance pesticides, which is generally improbable. Outrageous perspectives likening IPM with "bug free" cultivating will turn out to be progressively minimized and more adjusted perspectives will win.

There is no motivation not to help IPM as characterized by the FAO International Code of Conduct on the Distribution and Use of Pesticides (Article 2): Integrated Pest Management (IPM) implies a vermin the board framework that, with regards to the related climate and the populace elements of the irritation species, uses every reasonable strategy and techniques in as viable a way as conceivable and maintains the pest populaces at levels underneath those causing financially unsatisfactory harm or misfortune (FAO,1967).

Consequently, IPM is the best mix of social, organic and synthetic estimates that gives the most financially savvy, naturally solid and socially satisfactory strategy for overseeing illnesses, bugs, weeds and different irritations.

IPM is an information serious supportable methodology for overseeing vermin by consolidating viable social, natural, synthetic, and actual instruments in a way that limits monetary, wellbeing, and ecological dangers with the assistance of irritation scouts. IPM depends intensely on information on nuisances and yield connection to pick the best blend of locally accessible bug the board instruments (Fig. 8.1). Consequently, IPM is certifiably not a solitary item that can be put away on racks like pesticide, and it doesn't depend on single strategy to tackle all our bug issues.

Vermin likewise co-advance and adjust rapidly to single control strategies through regular choice, and that numerous techniques utilized at the same time, or an "incorporated" approach, is the best for long haul, practical administration programs.

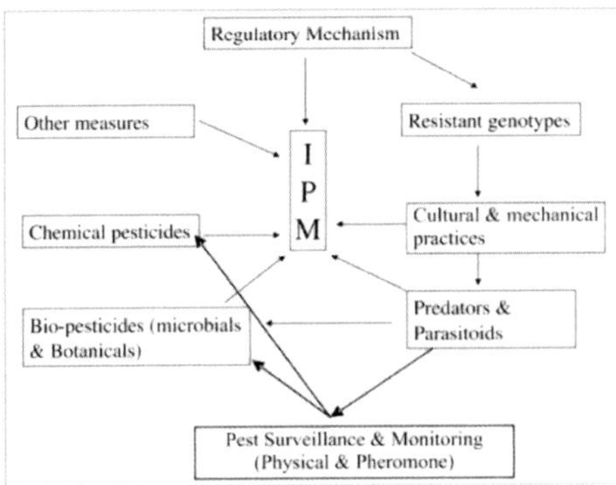

Figure 8.1. Dynamic sketch of IPM system

Source: Image by Academia.edu

IPM is neither organic nor it rely solely on biological control to achieve the desired sustainable outcomes. It does often try to assist and augment the effectiveness of natural enemies by limiting the impact of pesticide on their populations and provide clean and safe niche. It seeks to conserve balance between the crop and the natural environment.

The World Bank policy also promotes the conservation of natural habitats, and enhancement of the environment for long-term sustainable development. In the IPM idea, utilization of pesticides includes a compromise between bug control and the dangers of antagonistic consequences for non-target life forms, like regular adversaries, pollinators, untamed life, and plants, defilement of soil and water. Observing Agro Eco System Analysis (AESA): IPM has been advancing throughout the very long time to address the pernicious effects of engineered substance pesticides on climate at last influencing the interests of the ranchers. In current IPM (FAO, 2002) accentuation is given to Agro Eco System Analysis (AESA) where ranchers take choices dependent on bigger scope of field perceptions.

The soundness of a plant is controlled by its current circumstance which incorporates actual variables (for example sun, downpour, wind and soil supplements) and natural elements (for example irritations, illnesses and weeds).

This load of variables can assume a part yet to be determined which exists between herbivore creepy crawlies and their normal adversaries. Understanding the complicated collaborations in a biological system can assume a basic part in bother the executives.

It's anything but a methodology, which can be profitably utilized likewise functionaries and ranchers to dissect field circumstances as to bothers, protectors, soil conditions, plant wellbeing, the impact of climatic components and their interrelationship for developing solid yield.

A particularly basic examination of the field circumstances will help in taking proper choice on administration practice. The fundamental segments of AESA are:

- Plant health at different stages.
- Built-in-compensation abilities of the plants.
- Pest and defender population dynamics.
- Soil conditions.
- Climatic factors.
- Farmers past experience.

8.3. MANAGEMENT PRACTICES FOR INSECT PESTS

Figure 8.2. Management practices for insect pests

Source: Image by Flickr

8.3.1. Cultural Practices

- Deep summer ploughing followed by fallowing helps in exposing resting stage of pests.
- Inter-cropping with legume reduces borer incidence. Maize-Soybean/Maize-Cowpea/ MaizeGreen gram are some of the good examples.
- Use of well decomposed farm yard manure (FYM) reduces termite attack.
- Plant spacing 75 cm x 18 cm in Fall/autumn and 60 cm x 18 cm in Rabi is recommended.
- Balanced use of fertilizers (NPK 120:60:40) kg/ha and supplement of micronutrient.
- Proper water management practices to avoid foot rot.

8.3.2. Genetic Management

Figure 8.3. Diversification by genetic management

Source: Image by Newswise

Use certified seeds of recommended verities having built-in mechanisms for resistance of local pest problems as given below:

8.3.3. Mechanical Practices

- Removal of dead hearts will help to reduce second generation infestation.
- Use of bird scarer prevents seed damage.
- Manual collection and destruction of white grub and chaffer beetle during adult emergence period reduces the pest population.

8.3.4. Biopesticides

Soil application of neem cake for control of nematode and chaffer beetle

8.3.5. Biological control

- Conservation of naturally occurring biocontrol agents such as Trichogrammachilonis Ishii., Cotesiaflavipes Cameron, Carabids, Coccinellids, Chrysoperla, spiders and wasps, etc. and by reducing chemical pesticides.
- Release of Trichogrammachilonis @ 1,60,000 /ha. on 7- and 15-days old crop and subsequently if required.

8.4. PEST MANAGEMENT

8.4.1. Major Insect Pests of Maize and Their Management

There are four major pests of maize prevalent in India. These are spotted stem borer [Chilopartellus (Swinhoe)], pink stem borer [Sesamia inferens Walker], shoot fly [Atherigona spp.] and fall armyworm [Spodoptera frugiperda (J. E. Smith)].

Integrated Pest Management (IPM) focuses on administration of vermin through a mix of strategies like synthetic, natural, new trimming framework, adjustment of social practices, utilization of safe assortments and through mechanical techniques. The significant insects and IPM is depicted here.

1. Spotted Stem borer, Chilopartellus (Swinhoe)

Highlights of Identification

Spotted stem borer (Chilopartellus) is the main bug vermin of during Fall/ autumn season causing yield misfortunes in the scope of 26-80%in diverse agro-climatic districts of India. Eggs are level, oval, yellowish and are laid in covering groups. The hatchling is rich pink to yellowish earthy colored with 4 lines of specked stripes along the back with rosy earthy colored head. The larval period is finished in around 14-28 days.

The forewing of grown-up female moth is earthy colored yellowish with hazier scale designs shaping longitudinal stripes. In male moths, rear wings are pale straw-shading, and in females, they are white. The existence cycle is finished in around 5 a month and a half.

Nature of Damage

Grown-up of moth of spotted stem borer favors 3-5 leaf stage maize for egg laying. Eggs are by and large laid on lower surface of leaves. Eggs incubate in 3-4 days and the recently brought forth hatchlings slither inside the leaf whorl and feed in gatherings.

At the point when the moved leaves of whorl spread out arrangement of pin openings and papery windows are apparent, which is the principal manifestations of spotted stem drill assault. Following seven days onwards, hatchlings move out of whorl and bore upwards the creating tail, frequently arriving at meristem.

When meristem is taken care of upon, the leaf whorl evaporates known as 'dead heart' and the plant generally passes on and regularly leads to turners.

Integrated Pest Management

- Collection and destruction of the stubbles.
- Deep summer ploughing (not recommended under conservation agriculture)
- Planting of Napier grass in the boundary as a trap crop.
- Inter-crop maize with cowpea in 2:1 ratio.
- Two releases of Trichogrammachilonis 8 cards/ha (1,50,000 parasitized eggs/ha) at 7 and 15 days after germination.
- Remove and destroy dead hearts
- When infestation crosses 10%, spray Chlorantraniliprole 18.5 SC @150ml/ha
- Natural Enemies
- Egg parasitoid: Trichogrammachilonis
- Larval parasitoid: Cotesiaflavipes

- Pupalparasitoid: Xanthopimlastemmator, Tetrastichushowardii
- Predators: Chrysoperlacarnea, coccinellid, spider, ear wig, dragon fly, preying mantid, Pentatomid bug, reduviid bug, robber fly, rove beetles, wasp, king crow.

2. Pink stem borer, Sesamia inferens Walker

Features of Identification

Pink stem borer is the main vermin during Rabi season (winter) causing yield misfortunes in the scope of 25.7-78.9%. Eggs are rich white, dot like laid in 2-4 longitudinal lines inside the sheath of 1-2 lower leaves. Hatchling of this irritation is light pink in shading with a purplish hint. The larval period fluctuates from 22-36 days.

The grown-up female is medium measured, bold straw hued moth coppery touched sparkling scales with earthy colored streaks. Forewings have three little dark dabs on the dorsal side and a halfway earthy colored strip. Rear wings are white in shading.

The receiving wires were pectinate in guys and filiform in female moths. The all-out life cycle shifts from 40-53 days.

Nature of Damage

On incubating, pink drill hatchlings feed in camouflage inside the leaf sheath in gatherings and feed on the epidermal layer of the leaf sheath ideally on initial three leaf sheaths. The hatchling drill into the focal shoot bringing about evaporating of developing point and arrangement of dead heart in youthful plants.

The hatchling structure round formed passages inside the stem and leave openings at the surface. The passages are loaded up with excreta. Now and then roundabout ring like cuts on lower entomb hubs will be seen due to larval feeding.

Management

- Collection and destruction of the stubbles
- Deep summer ploughing (not recommended under conservation agriculture)
- Removal and destruction of dead hearts.

- When infestation crosses 10%, spray chlorantraniliprole 18.5 SC @150ml/ha
- Natural Enemies
- Egg parasitoid: Trichogrammachilonis
- Larval parasitoid: Cotesiaflavipes
- Predators: Chrysoperlacarnea, coccinellid, spider, ear wig, dragon fly, preying mantid, Pentatomid bug, reduviid bug, robber fly, rove beetles, wasp, king crow.

3. Fall Armyworm, Spodoptera frugiperda (J. E. Smith)

Highlights of Identification

Fall armyworm (FAW) is an obtrusive creepy crawly bug making genuine harm maize at all phases of its growth. It was first announced in quite a while in May, 2018. The female moths lay egg masses on upper or under side of the leaf and covered with tan shaded scales.

Each egg mass contains 50-150 eggs. Hatching period changes from 4-5 days. The hatchlings are smooth-cleaned and differ in shading from light tan or green to dull dim body with three rich yellow dorsal and sidelong lines. The hatchling contains rosy earthy colored head with dominating white, altered Y-formed stitch between the eyes.

The hatchling contains 6 instars in which the larval period changes from 15-18 days. On eighth and ninth stomach fragment, four enormous spots are organized in a square shape on section 8 and in trapezoid design on portion 9. Pupa is ruddy earthy colored in shading.

Following 7-9 days, grown-ups rise up out of pupae. The forewing of grown-up male moth comprises of grovel shaded spot and a white fix at the apical edge of the wing while female moths are less unmistakably stamped going from uniform grayish earthy colored to a fine mottling of dim and earthy colored.

The absolute lifecycle finishes in around 30-35 days which fluctuate as per climatic conditions. Grown-up life span shifts from 4-7 days. The grown-up moth can fly up to 500 km before ovi-position.

Nature of Damage

FAW assaults all phases of maize crop from seedling rise (V2) to ear advancement (R6). The youthful hatchlings of FAW feed in and around the whorl leaves by scratching and skeletonizing the upper epidermis leaving a brilliant straightforward layer coming about into papery spots.

The harm likewise brings about pinhole indications on the leaves. More seasoned hatchlings remain and feed inside the whorl. The harms by late instars (fourth instar onwards) bring about broad defoliation of leaves and presence of a lot of fecal pellets in whorls.

Harm during vegetative stage prompts leaf harm however on the off chance that harm occurs during conceptive stage it might harm tufts or may exhaust inside the corn ear and destroy the portions. The whorl harm by FAW brings about critical yield misfortunes while ear taking care of results in both quality and yield decrease.

Flowering to harvest

Hand picking and annihilation of larvae boring into ears:

At 10% ear damage, application of Bacillus thuringiensis v. kurstaki formulations (400g/acre) @ 2g/litre or Metarhiziumanisopliae or Beauveria bassiana with spore count of1×108cfu/g (1 kg/acre) @ 5g/litre or SfNPV (600ml/acre)@3ml/ litre or entomopathogenic nematode (EPN) (4kg/acre) @20g/litre of water is recommended.

#Pheromone traps- Funnel trap with FAW draw ought to be introduced at a tallness changed every week coordinating with crop shelter. Traps ought to be isolated by a base distance of 75 feet. Notice snares for number of moths got twice or once in a week and work out the catch/day. FAW draws ought to be changed once in 30 days if there should be an occurrence of observing.

Readiness of Neem Seed Kernel Extract (NSKE) for one section of land 10 kg of neem seed bit is needed for one section of land. Crush 10 kg of need seed portions to make powder. Absorb the powder 50 liters of water overnight. Stir and channel the substance utilizing cotton material. Add 200 g cleanser powder or 200 ml of cleanser answer for the sifted arrangement. Make up the volume to 200 liters by adding water.

Caution upon release of egg parasitoids–Minimum multi week span ought to be there between parasitoid delivery and use of neem or synthetic insect sprays

Precautionary measures for pesticide use: Not in excess of two synthetic showers are to be utilized in whole harvest span. Same compound should not be picked for second shower. Splashes ought to consistently be coordinated towards whorl and applied either in early hours of the day or in the evening time.

Utilize defensive dress, facemask and gloves during arrangement and utilization of pesticides. Enter the field just a brief time in the wake of splashing pesticide. Interval between use of compound insect spray and collect of corn ought to be least 30 days.

Natural Enemies

- Egg Parasitoid: Trichogramma pretiosum, Telenomusremus
- Egg-larval Parasitoid: Chelonussp
- Larval parasiotoid: Coccygidiumsp, Campoletischloridae
- Predators: Predatory Pentatomid bugs, ear wigs
- Entomopathogen: Nomuraearileyi

4. Shoot Fly (Atherigonasoccata Rondani; Atherigonanaqvii Steyskal)

Features of Identification

Shoot fly, Atherigona spp. was accounted for to be a genuine irritation in spring maize which cause substantial plant misfortune whenever planted during third seven day stretch of February to first seven day stretch of March.

Spring maize is developed in North-Western plain zone of India in the provinces of Punjab, Haryana, Uttarakhand and Uttar Pradesh. It was answered to cause up to 60% plant misfortune and around 21% grain yield misfortune.

The types of Atherigona present in Punjab is A. naqvii while, the locale coterminous of Haryana, Delhi, Uttarakhand and Uttar Pradesh has an animal types complex with A. soccata as well as A. orientalisin larger part and A. naqvii in minority, the extent of which is yet to be settled.

Eggs of A. soccata and A. orientalis are prolonged like a little rice grain, smooth white in shading with two wing-like projections and those of A.

naqvii are round and hollow with fine edges on surface. The incubation period is 1-3 days. Larval period is 7-10 days with 3-4 instars.

Full grown maggot is yellow in color. Pupation takes place inside the stem. The pupa is dark brown, barrel-shaped. The pupal period lasts for about a week. The adult lives for 3-4 days. The life cycle is completed in about 3 weeks.

Nature of Damage

Atherigona spp. invades maize from multi day old seedlings to three weeks after rise by ovipositing of eggs on the abaxial surface of basal leaves, shoot and in soil near base of the plant. Maggots drill into shoot while taking care of, progressively killing the developing point prompting shriveling of focal shoot called dead heart, which is framed inside about fourteen days of germination.

Management strategies

- Sowing must be completed by first fortnight of February
- Seed treatment with imidacloprid 600 FS @ 6 ml/kg seed or Thiamethoxam 30 FS @ 8.0 ml per kg seed
- Removal and destruction of dead hearts
- Natural enemies
- Predators: Coccinellid, spider, robber fly, Pentatomid bug, ear wigs.
- Minor pests of maize
- Cob borer [Helicoverpa armigera (Hubner)]

Features of Identification

Helicoverpa armigera is an arising bug of maize, particularly in sweet corn. Eggs are circular fit and velvety white in shading, laid separately on silk. There are five or six larval stages. The hatchling shows shading variety from greenish to brown.

It has dulled earthy colored dim lines on the body with horizontal white lines. Pupa is earthy colored, adjusted both anteriorly and posteriorly, with two tightening equal spines at back tip. Forewings of grown-up moth contain line of seven to eight blackish spots on the edge and a dark comma-molded checking in the center underside of every forewing.

There was a recognized shading design among male and female moths. The forewings of guys were greenish-dim in shading which later blurred to straw yellow while the rear wings are cream to light yellow with dull earthy colored external peripheral band.

If there should arise an occurrence of females, the shade of forewing was light brown while the rear wings were cream to light yellow with articulated dim earthy colored groups. Females were likewise recognized by the presence of tuft of hairs on the tip of mid-region.

Management strategies

- Cultivate sweet corn hybrids with husks well covering the ear-tip.
- Hand pick and destroy larvae before it enters inside the ear
- Installation of pheromone traps @4/ha for monitoring purpose
- Application of BtK formulations @ 2g/l or HaNPV @500LE/ha directed into ears if heavy infestation is noticed.

Natural enemies

- Egg parasitoid: Trichogramma chilonis
- Larval parasitoid: Campoletischloridae, Braconsp
- Predators: Chrysopids, Anthocoreids, coccinellids, pentatomid bugs, wasps, spiders
- Entomopathogens Beauveriasp, Metarrhiziumsp
- Tobacco caterpillar [Spodoptera litura (Fabricius)]

5. Tobacco caterpillar [Spodoptera litura (Fabricius)]

Features of Identification

Spodoptera species are appropriated all through India. Females lay eggs in masses of around 200 to 300 on the underside of leaves and covered with earthy colored scales. Eggs are laid in 2-3 layers. The hatchling is variable in shading, sides of body with dim and light longitudinal groups; dorsal side with two dull semilunar spots along the side on each portion, aside from the prothorax; spots on the first and eighth stomach sections bigger than others, intruding on the horizontal lines on the principal fragment.

A splendid yellow stripe along the length of the dorsal surface is normal for S. litura hatchling.

The hatchlings feed in bunch when they are youthful however fanned out as hatchlings develop. Grown-up hatchlings tumbled from plants pupate in a little cell in the dirt/cobs. In guys, forewing contains reniform earthy colored spot, lengthen orbicular spot, white fork in the middle region, a column of dull earthy colored glass markings along external edge and an enormous yellowish or light earthy colored fix on the middle region contiguous inward edge.

The forewing of female moth is dim to ruddy earthy colored with an unequivocally variegated example and paler lines along the veins. The rear wing is grayish white with dark edges. The lifecycle requires around 25-30 days.

Nature of Damage

On incubating, the hatchlings feed on delicate leaves in gatherings. They scratch the surface yet don't really puncture it's anything but, a window sheet impact. Under extreme pervasion, the whole youthful plant might be consumed. Later on, they move and feed on the leaves which gives slender papery appearance.

8.5. GENERAL GUIDELINES FOR MANAGEMENT OF RESISTANCE

The general guidelines if adopted can prevent development of resistance by various pests in most of the agricultural situations. The general approaches to avoid them are as follows:

- Insecticides
- Maintain good plant health,
- Delay the spray of insecticide as far as possible.
- Monitor populations and use economic thresholds
- Use all available tactics for management of a particular arthropod (insect or mite)
- Limit selection pressure throughout the season and remember spraying for one pest may influences another
- Limit use of one chemical molecule at a time and rotate chemical molecule and/or modes of action, and
- Use appropriate rates

8.5.1. Fungicides

- Try not to develop huge spaces of profoundly powerless assortments in endemic regions. Safe assortments ought to be utilized to decrease dependence on synthetic pesticides.
- Make full utilization of non-fungicidal control measures e.g., discard crop flotsam and jetsam and control security and substitute host, which harbor sickness.
- Monitor crops routinely for sickness and treat before the contamination gets set up.
- Use fungicides just in the unavoidable circumstances where the danger of sickness warrants therapy. Utilize compelling fungicides with various methods of activity as substitute splashes. Combinations of eradicant fungicides with protectants materials offer the most adaptability just as diminishing opposition hazard.
- While defining splash programs, consider any previous utilization of fungicides bunches as seed treatment.
- Do not surpass the most extreme prescribed quantities of utilizations to each edit for a specific fungicide bunch. Stay away from rehashed uses of extremely low portions.

8.6. SAFETY PARAMETERS IN PESTICIDES USAGE

Safety parameters interalia grouping of harmfulness according to Insecticides Rules, 1971, WHO characterization of dangers, shade of toxicity triangle, First guide measures, side effects of harming and treatment of harming, the augmentation functionaries of the State Department of Agriculture need to utilize this data as under:-

- Basic precautionary measures which are needed to be taken according to arrangement of harmfulness just as peril standards by WHO might be viewed according to Annexure – IX.
- The augmentation functionaries are to instruct the ranchers on wellbeing utilization of pesticides with the assistance of shading harmfulness triangle as the cultivating local area can follow the shading and comparing security safeguards.
- The side effects of harming should be known to the augmentation functionaries to empower them to stretch out medical aid measures to influenced people to the degree conceivable.

- Basically, the data on medical aid measures and therapy of harming is needed to be given by the augmentation functionaries to the specialists at Primary Health Centers just as to Private Doctors nearby showering of pesticides.
- Extension functionaries should guarantee that names of normal pesticides during plant insurance gauges alongside a duplicate of the pamphlet which is a fundamental piece of a pesticide compartment should be made accessible to the specialists nearby plant assurance activities.
- Extension functionaries are to demand the specialists to mediate in acquirement of counteractants for various pesticides as referred to under "Treatment of harming".

8.7. INSECT PEST MANAGEMENT IN MAIZE

Figure 8.4. Stem borers hidden in maize tassel

Source: Image by Flickr

Maize can be assaulted by a wide scope of creepy crawlies. The primary nuisances of maize are helicoverpa and various soil creepy crawlies. Minor nuisances happen sporadically and won't be an issue each season.

Maize pests by crop stage

Vermin can happen at least one development stages (Insects that can conceivably be a significant bug are demonstrated in intense):

8.7.1. Helicoverpa armigera (Corn Earworm)

Helicoverpa armigera is major, broad, ordinary pest. Female moths lay eggs on the stem, leaves (the two sides) tufts, silks and husks on the upper 66% of plants. Caterpillars incubating before silking because little harm to tufts however may cause harm when relocating to cobs. Hatchlings from eggs lay on silks or husks may cause critical harm.

Silk harm diminishes fertilization and grain-set. Taking care of harm additionally happens on the best 1-3 cm of the cob, and may bring about the presence of mycotoxins.

Leaf harm can demonstrate bug presence. Equal lines of openings are indications of benefiting from unopened leaves.

Helicoverpa are typically not considered prudent to control, besides in high worth seed maize. Note that helicoverpa hatchlings might be mistaken for armyworms or cutworms.

Maize assortments with husks expanding 50-80 mm past the highest point of the cob and shutting firmly around the silks limit the section of hatchlings into the cob. Watering during dry climate keeps the husks from extricating.

Maize crops frequently have undeniable degrees of advantageous insects (hunters and parasitoids) that might be hurt by bug spray applications. The joined activity of normal foes (counting hunters of eggs, hatchlings and pupae, parasites of eggs and hatchlings, and caterpillar illnesses) can have a critical effect. Development to a dept of 100 mm obliterates overwintering pupae.

Substance control should target little caterpillars (up to 5 mm) and be aimed at decorations and arising silks. The best item to use in an incorporated vermin the executives framework is a normally happening nucleopolyhedo virus (NPV). There are various financially defined NPV items available for the control of helicoverpa. For more data, visit the NPV page.

8.7.2. Soil Insects

Soil-staying bug pests can truly decrease plant foundation, plant populaces, plant development, and resulting yield potential, and ought to be observed for preceding plating. Seed medicines hclp hinder taking care of. In-wrinkle splashing ensures youthful roots and shoots.

Figure 8.5. Soil bugs

Source: Image by NPR

Press wheels can diminish harm from bogus wireworm hatchlings and earwigs by empowering plant rise and firming the dirt to lessen creepy crawlies' capacity to travel through it. Soil teasing may decrease harm by dark field earwigs and crickets that assault the tips of creating prop-roots (auxiliary roots). Shallow planting in warm wet soil will energize quick harvest rise and development, accordingly diminishing the effect of creepy crawlies. For general data about soil insects and germinating seed baits (GSB) perceive How to screen for and perceive soil insects.

Dark field earwig is an inconsistent and possibly significant bug of maize, dark earwigs eats recently planted and developing seed and the foundations of yields bringing about helpless foundation. Benefiting from optional roots may make the plants fall over as they get bigger.

Genuine harm is typically bound to soils that hold dampness well, and earwigs incline toward developed soils to undisturbed soil (zero till). Note: savage earwigs are typically bigger and light earthy colored in shading.

Screen crops in the wake of planting until foundation. Burrow and sifter soil is required to recognize grown-ups and sprites preceding planting. Use sprouting seed snares and control if in excess of 50 earwigs in 20 developing seed draws. Grain draws containing insect spray applied at planting offer best security. Insect spray seed dressings give some security. In-wrinkle splashes are not viable in securing against thick populaces. Use press wheels at planting.

Genuine wireworm hatchlings drill into growing seed and bite on seedling roots and shoots bringing about diminished life, shrinking or seedling demise.

Use growing seed traps or soil testing to recognize hatchlings before planting. Screen crops in the wake of planting until foundation. One hatchlings/sprouting seed lure warrants control. Seed dressings, in-wrinkle splashes and granular insect poisons offer some control.

False wireworm larvae attack developing seeds and seedling roots and shoots in spring, bringing about sketchy stands. Harm is generally normal in early planted yields with low harvest buildup (for example developed enclosures). Grown-ups may harm summer seedlings by gnawing at or over the ground level and replanting might be required.

To recognize, either hand filter 10 soil tests (30 x 30 cm) or spot 10 sprouting seed traps all through the enclosure. One hatchling for every example warrants control. Get ready ground for even and quick germination. Utilization of press wheels at planting gives some control. For hatchlings, use seed medicines or in-wrinkle splashes.

For grown-ups, utilize broke grain lures. Regular adversaries give little control. Infestations detected after crop rise can't be controlled. Cutworm hatchlings feed on leaves and stems of youthful plants, and 'slice' down plants to eat the leaves. Incomplete harm to stems may make the plant wither. Hatchlings regularly cover in the dirt during the day and twist into a 'C' shape when upset.

Cutworms might be found in any dirt sort and frequently move into the yield from abutting wall lines, fields or weedy fallows. Harvest regions assaulted by cutworms will in general be inconsistent and the most elevated danger period is during summer and spring.

Assess arising seedlings two times seven days. Treat seedlings when there is a quickly expanding space of pervasion or extent of yield harm (>10% seedling misfortune). Treat more seasoned plants if over 90% of plants are invaded or over half of plants have 75% or more leaf tissue misfortune. Spot medicines (for example along field edges) might be effective. Shower when caterpillars are taking care of (sunset evening). Keep fallows clean and kill weeds from enclosure borders for at any rate one month prior to planting. Serious harm to arising yield can happen when huge hatchlings are compelled to move from weed has into the harvest following splashing of the weeds. Cutworms are assaulted by a scope of regular foes, for example, parasitoids, hunters and infections.

8.7.3. Sucking Pests

Maize leafhoppers and maize thrips are boundless yet sporadic in Queensland, and can quickly re-overrun cops in the wake of showering meaning more than one splash might be required.

Maize leafhoppers are a pest pest of developing maize. Generally normal during pre-fall, they suck sap and high populaces (>15/plant) can send wallaby ear mycoplasma. Mycoplasma-tainted plants are dim green tone with thickened veins on the underside of leaves. Crossover assortments offer some protection from wallaby ear. Review crops week by week during the vegetative stage, and control if in excess of 10 leafhoppers/plant and wallaby ear side effects are available.

Maize thrips attack seedling and vegetative development stages. Thrips in the whorl can stop the development of little plants. Contaminated plants may have a yellow or gleaming patches on the leaves and a dried up or withered appearance. Harm is more extreme if plants are focused by dry season or water logging. Investigate week after week during seedling and vegetative stages, and control if thrips are found in the throat joined with yellowing of the throat or necrotic stripes on youthful leaves.

Corn aphid is the most well-known aphid species on maize and can influence any yield stage. Grown-ups and sprites suck sap and produce honeydew. High numbers can make plants become yellow and seem unthrifty. Yield misfortune may happen on water focused on plants.

The frequency of harm is for the most part too low to even think about justifying control. Substance control alternatives are for the most part not financially savvy and the insect poisons that control aphids may adversely affect normal adversaries. Assess at week after week spans. Hunters of aphids incorporate ladybird larvae, lady bugs, big - eyed bugs, larvae of green lacewings and larvae of hoverflies. Wasp parasitoids preserve and kill aphids. Green vegetable bugs (GVB) are widespread but irregular pests of maize during summer. Adults and nymphs feed by piercing and sucking on developing cobs, and may severely deform cobs. Although chemical control may be cost effective, there are currently no threshold levels for GVB in maize.

Two-spotted parasite is a far reaching however unpredictable pest of maize during seed fill to development. Vermin are generally present towards the finish of the yield cycle during pre-fall/harvest time and are supported by sweltering, dry climate. Grown-ups and fairies puncture and suck on lower leaf surfaces, causing yellowing on the upper leaf surfaces.

Fine webbing on the lower leaf surface shows their essence, and hefty pervasions will bring about leaf parching, leaf drop and yield misfortune. Starting invasions can be sketchy. No edges are accessible for insects in maize and control isn't savvy. Moderate regular adversaries through decrease of wide range bug sprays against different nuisances as the disturbance of normal foes can erupt parasite flare-ups.

Red banded safeguard bug is a boundless however minor summer/ pre-winter bother that can be mistaken for green vegetable bug. Grown-ups and sprites feed by puncturing and sucking on creating cobs, causing disfigurements, like green vegetable bug harm.

8.7.4. Chewing Pests

Figure 8.6. Moles by bug

Source: Image by impakter

Grasshoppers are irregular and conceivably significant irritations of maize. Grown-ups and containers bite sporadic pieces from leaves and stems and can cause total defoliation short-term if populaces are sufficiently high. Species found in Queensland include: Australian plague grasshoppers,

Migratory beetles, and Spur-throated insects. The Australian Plague Locust Commission gives subtleties of container relocations.

White fringed weevil larvae chew into lateral roots causing death and reduced vigor. Infestations are usually patchy. Mass emergence of adults occurs after rain in November-January, and damage is often worse when two host crops (e.g., maize, peanuts, chickpea, lucerne) are grown in sequence.

Leaf creepy crawlies are a sporadic pest of maize:

The red shouldered leaf scarab is confined to seaside regions and can plague at any phase of yield development. Multitudes of grown-up insects move into a yield and feed on foliage, decorations, silks and the husk at the highest point of the cob.

Injury to silks may diminish seed set. Taking care of opens the cobs to different insects and illnesses. Invasions will in general be sketchy so exhaustive week after week checking is required. Control is justified if 95% of plants in a space are pervaded and 70% of banner leaves are eaten. Compound control is practical at high invasions.

Amassing leaf scarabs are a minor, sporadic irritation in Queensland. The dark leaf creepy crawly can be a significant irritation in northern regions. Leaf creepy crawlies are probably going to be available during the seedling stage.

Hatchlings feed on attaches causing up to 40% seedling passing. There are at present no suggested techniques for observing and financial injury levels have not yet been set up. Try not to plant maize following grass field.

Armyworms can happen in enormous numbers particularly when great downpour follows a dry period. During the day the normal armyworm covers in the throats of plants or in the dirt and arise after dusk to take care of. Day feeding armyworm is dynamic during the day.

Youthful plants might be defoliated or killed. More seasoned plants can grow out of harm however seed yield might be decreased. Indications of harm incorporate bit leaf edges and fecal pellets at the foundation of youthful plants or in the throats of more established plants.

Screen during seedling and vegetative stages. Egg lays are regularly connected with hefty precipitation so check for hatchlings half a month after precipitation occasions. Search for hatchlings under hunks of soil, under vegetation and at the foundation of plants. As an aide, control is justified if out of a tally of 30 plants, 27 are pervaded, and more than 21 have in any event 75% banner leaf misfortune. Numerous synthetics will control armyworms however their adequacy is regularly reliant upon great entrance and control might be more troublesome in high-yielding thick shade crops, especially when hatchlings are resting under leaf litter at the foundation of plants.

As hatchlings are generally dynamic around evening time, showering in the early evening or evening may create the best outcomes. Armyworm hatchlings are assaulted by various parasitoids that may help with diminishing the power of episodes, in spite of the fact that are probably not going to give ideal control if armyworm numbers are high. Hunters incorporate green carab scarabs, savage safeguard insects and maybe normal earthy colored earwigs. Viral and parasitic infections are recorded as causing mortality of armyworms.

Yellow peach moth is a minor and irregular pest of maize in Queensland. Larvae tunnel into stems or cobs producing masses of webbing and excreta at the tunnel entrance. There are no chemical controls available. The habit of mining into stems and cobs makes spray application ineffective as larvae are not exposed to insecticide. They are similar in appearance to pink-spotted bollworm larvae.

8.8. CONCLUSION

This chapter provides an insight to the reader by highlighting the integrated pest management practices in the cultivation of maize. It begins with describing the different management practices for insect pests which involves the cultural practices, genetic practices, mechanical practices, bio pesticides and biological control. In the second half, it addresses the pest management by illustrating the major insect pests of maize and their management, along with the feature of identification by giving highlights of identification, nature of damage. It goes on discussing the pre-planting practices, and talks about it in detail. Then it outlines the flowering used to harvest, the natural enemies in cultivation which are egg parasitoid, egg-larval parasitoid, different predators, shoot fly etc.

It also throws light on the management strategies and programs, together with mentioning the general guidelines for management of resistance which includes insecticides and fungicides and the corn earworm which is in biological term called *helicoverpa armigera*. In the end, it states the safety parameters in pesticide usage and the various forms of insect pest management in maize.

REFERENCES

1. ICAR, 2021. Pest Management – ICAR-Indian Institute of Maize Research. [online] Iimr.icar.gov.in. Available at: <https://iimr.icar.gov.in/pest-management-2/> [Accessed 30 June 2021].
2. Sangit Kumar, Pradyumn Kumar, Jugal Kishor Bana, M Shekhar, S N Sushil, A K Sinha, Ram Asre, K S Kapoor, O P Sharma, S Bhagat, M Sehgal, T Boopathi, N Amaresan, C Chattopadhyay, K Satyagopal and P Jeyakumar, 2014. INTEGRATED PEST MANAGEMENT PACKAGE FOR MAIZE. [ebook] National Centre for Integrated Pest Management. Available at: <https://niphm.gov.in/IPMPackages/Maize.pdf> [Accessed 30 June 2021].

Chapter 9

Advanced Maize Production Technologies

CONTENTS

9.1. Introduction..192
9.2. Improved Technologies for Higher Maize Production......................196
9.3. Improved Production Technologies...198
9.4. Other New Technologies Available Today.......................................209
9.5. Production technology of maize..213
9.6. Conclusion...215
References...217

This chapter sheds light on the improved and advanced technologies which are implemented for higher maize production. It begins with crop establishment which includes sowing time, seed rate, seed treatment, crop geometry. As the chapter proceeds it deals with the methods of planting like the zero tillage, raised bed planting, flat sowing, and transplanting.

The second half of the chapter illustrates the water management which incorporates flood irrigation, furrow irrigation, above ground drip irrigation, sub-surface drip irrigation, partial root dying irritation.

Then it highlights another kind of management which is the weed management done through different process like the non-chemical control which includes manual weeding and mulching along with this it also outlines the nutrient management which constitutes of integrated nutrient management, fertigation etc.

It then throws light on the insect and pest management which is has different aspects and deals with integrated pest management, biological pest management. Towards the end, it discusses the different new technologies that are available in recent years such as the data management technologies, production technologies and the manual strategies.

9.1. INTRODUCTION

Maize is the most versatile crop which is used as food, feed, and fodder and in recent past as a source of bio-fuel. It has wide adaptability. Worldwide it is being cultivated in over 170 countries representing an area of 185 million ha with a productivity of 5.62 t/ha (FAOSTAT, 2017).

Out of world maize production of 1037 million MT, SAARC countries comprising of Afghanistan, Bangladesh, Bhutan, India, Maldives, Nepal, Pakistan and Sri Lanka represent 3.2% with a productivity of 3.8 t/ha. Among SAARC countries, India represents the highest area (78.9%) followed by Pakistan (9.7%), Nepal (7.9%), Bangladesh (2.6%) and Sri Lanka (0.6%).

The highest productivity of 6.9 t/ha is reported in Bangladesh followed by Pakistan (4.3 t/ha), Sri Lanka (3.4 t/ha), Bhutan (3.0 t/ha), India and Nepal (2.5 t/ha). Duringthe decade covering 2005 and 2014, maize region and usefulness in SAARC nations has expanded at a yearly accumulate development rate (ACGR) of 2.2% and 5.2% per annum, separately.

The amazing advancement of maize in SAARC nations is an aftereffect of expanded interest in poultry and other feed industry as opposed to increment popular for food. Maize R&D endeavors in a portion of the SAARC nations are quite a few years old while in some like in Bangladesh, maize development is a greater amount of late presentation. Since these nations are in nearness, the difficulties like biotic or abiotic stresses are moderately comparative. The requirement for innovative intercessions may likewise have similarities.

Then again, SAARC nations can contend all the more enthusiastically in trade advertises and guarantee benefits for a wide scope of entertainers in the worth chain of the maize crop. This gives openings and difficulties to working in association with the CGIAR, the private area, strategy creators and different partners to improve proficiency and effect.

Maize is high yielding; stress open minded and nutritious. Maintainable Intensification of Maize-based Cropping Systems adding an incentive for maize makers, processors, and shoppers in South Asia would be an extraordinary financial incentive for the improvement of business of the ranchers.

It has been shown that the maize is an arising high-esteem crop, having the most elevated normal ranch yields in Asia. Monetarily, half breed maize is considerably more beneficial than Boro rice, wheat, or most other contending Rabi crops.

Despite the fact that the harvest can be become both in fall/autumn and in rabi seasons, yet the probability of acknowledging extremely high return is conceivable just during the rabi season, particularly in Bangladesh. Maize development is relied upon to extend driven by poultry, fish and steers feed industry.

It is currently primarily filled in north-western, south-western regions and focal districts of Bangladesh and the production production is designated generally for poultry, fish feed and animals which made this area defenseless. An arrangement of new innovations for practical and concentrated maize production production frameworks is arising in Bangladesh of which some are being advanced and embraced rapidly.

The manageability of half and half maize production production in Bangladesh relies upon the utilization of value crossover seeds, great agronomic administration, use of offset manures alongside soil richness preservation, and other administration rehearses.

Advances in sub-atomic advances and methodologies, combined with new instruments for abiotic stress phenotyping, will give an incredible mix to creating improved pressure versatile maize for stress-inclined agro-ecologies of the country as their seriousness and recurrence increments. In Bhutan, maize is developed by more than 36,000 families for food and pay.

Maize has a bigger region under development when contrasted with rice and is likewise created in bigger amounts. The maize esteem chain shows that it's anything but a means ware with as much as 94% being held at home for utilization.

Other than the maize grain, maize stover has various utilizations in their blended cultivating framework, from feed to bedding material for dairy cattle and eventually adding to natural compost. Along these lines, the food security part of maize can't be downplayed. Notwithstanding, the pattern is changing and with higher urbanization and higher salaries, inclination for rice is expanding.

Along these lines, additional time, the significance of maize as a staple is probably going to diminish. Maize has a mechanical esteem and can be formed into a potential money crop for the ranchers in the event that it very well may be created in satisfactory amounts.

The venture on maize ware program in the twelfth FYP should endeavor to expand production production with the extension to produce sufficient volume that can support maize utilizing undertakings. Moreover, the example of overcoming adversity of maize in India and other SAARC nations over most recent twenty years has stayed remarkable.

The production production has expanded both because of huge expansion in region just as efficiency. Hereditary addition because of presentation of mixtures, all the more exactly single cross half breeds, and better harvest the board have contributed towards usefulness improvement. The drive behind the maize upheaval in India and other SAARC nations is the interest of maize grain as feed, all the more especially poultry feed. Despite the fact that amazing indications of progress have been made a colossal hole among potential and acknowledged yield exists, which can be spanned uniquely through up-scaling and out-scaling the demonstrated advancements among the maize ranchers.

The maize R&D policy needs to follow a bottom-up approach with involvement of all stakeholders. Changing climate has brought a new dimension to the maize R&D scenario. In any case, maize accepts preferable benefits over different cereals in tending to this test.

India being deliberately situated with its long history of coordinated maize examination may assume a significant part to lead the maize R&D in

SAARC by sharing skill, granting preparing and going about as seed center for the entire locale. Government policies encouraging closer cooperation among SAARC nations will profit not one country but rather all others too.

On account of Nepal, maize has been developed from Terai to mountain districts of Nepal. It is significant in food and healthful security in mid and high slopes; and feed and feed security in Terai locale of Nepal. New maize innovations are being created across the globe and accomplishments are being shared through CIMMYT and other CGIAR accomplice associations.

New advancements recognized, created and confirmed by investigates have been advanced by instructive and expansion frameworks. Due to similitudes in agro-climatic conditions among SAARC nations, the advancements recognized as promising in one nation ought to be divided between one another to assist the district. New technologies viz. hybrids, QPM, pro-vitamin A, resource conservation technologies; rational and scientific uses of inputs should be promoted aggressively.

In Pakistan, the planting/water conservation techniques being used in maize include ridge planting, bed planting, flat with line planting and broadcast planting. Nonetheless, Ridge Planting (RP) is the best technique for getting most extreme grain yield per unit region.

The boundaries to be considered for getting most extreme grain yield in maize incorporate plant stature (cm), number of leaves per plant, leaf rolling, leaf region per plant (cm^2), ear tallness (cm), days to half tasseling, days to half silking, grain columns per ear, grains per line, ear length and size (cm), 100 grain weight (g), stem size at fourth internode, shelling rate and grain yield (kg/ha).

Maize is second significant grain as far as development degree and production in Sri Lanka. In any case, 80-85% of production utilized for poultry and steers feed industry. The maize is given as the primary goal crop under other field crops in innovative work programs at Field Crops Research and Development Institute of DoA. The few OPVs and cross breeds were created in the country in a joint effort with CIMMYT during most recent 40 years.

Ranchers are requesting cross breed maize seeds and 95% of maize region is under mixture maize. The 95% of the absolute half breed seed prerequisite is met by imported high yielding mixtures. Subsequently public normal usefulness has expanded up to 3.6 t/ha.

The primary disadvantage is supporting the usefulness in maize arrives because of land corruption and other abiotic and biotic burdens. Subsequently, SAARC Member States ought to be viewed as independent strategy in maize production by expanding efficiency through various vital plans.

The future focal point of examination is advancement of half and half maize with high return potential and giving exceptional accentuation on pressure flexibility. Therefore, in the midst of the relative multitude of difficulties and difficulties, cooperation and trade of thoughts, innovation and vision among SAARC nations can bring food, feed and wholesome security, which is significantly contribute the lessening neediness.

9.2. IMPROVED TECHNOLOGIES FOR HIGHER MAIZE PRODUCTION

Figure 9.1. Climate Smart crop production practices

Source: Image by FAOUN

A variety of production advances, from land planning to collecting, has been suggested for maize crop. Being non-tillage crop, ideal plant populace can be accomplished if reasonable yield foundation procedures like technique for planting, planting time, seed rate, seed treatment, crop calculation and so forth, are followed.

Weeds can be managed well either by two hoeing 15–30 days after sowing or herbicides like atrataf 50 WP (atrazine) at 2 kg/ha on medium to heavy textured soils and 1.25 kg/ha in light soils within 10 days of sowing, using 500 liters as pre-emergence or spray 262.5 ml/acre Laudis 420 SC (tembotrione) in 375 liters of water at 20 days after sowing.

Integrated nutrient management strategy renders use of farm yard manure at 10–15 t/ha, Paddy straw compost at 450 kg/ha or synthetic fertilizers at 120 kg N, 60 kg P2O5 and 40 kg K2O per hectare for hybrids and 80 kg N, 30 kg P2O5 and 20 kg K2O per hectare for composites.

Incorporated pests management approach stresses on utilization of physical, synthetic or organic measures for the control of creepy crawly bugs. Maize drill can be constrained by showering coragen 18.5 SC at 75 ml utilizing 150 liters water/ha.

Drying of maize produce should be possible sun drying, smoking or air drying for bringing better market price. Maize is known as the Queen of Cereals' because of its interest and more extensive versatility. It is the second most significant oat crop on the planet as far as grounds and creation. Worldwide production of Maize was around 1040 million MT in the year 2016–2017, where in USA and China contributed around 38 and 23%, separately.

In India, maize is the third most significant food crop after rice and wheat, where around 15 million ranchers are occupied with maize development. In India, Andhra Pradesh positions first in maize production followed by Karnataka with percent portion of 20.9 and 16.5, individually [2]. It's anything but a portion of 9% in about Rs. 100 billion farming area total national output.

Maize can be developed effectively in loamy sand to substantial dirt, all around circulated air through, unbiased pH soils. As of tropical beginning, it is exceptionally delicate to water stagnation, so keep away from the development in low-lying or helpless waste fields.

Besides, broadened low temperature less 5°C severally influences the harvest. Ideal scope of temperature for better harvest development and yield acknowledgment is 25–35°C [4]. Being day nonpartisan, maize harvest can be developed during the time which prompts high return levels in a brief timeframe. In this section, we will talk about a variety of various production advancements to be trailed by ranchers for fruitful development and better acknowledgment of yields.

9.3. IMPROVED PRODUCTION TECHNOLOGIES

9.3.1. Crop Establishment

Selection of Cultivar

Type of cultivar/hybrid to be grown depends on the crop season namely, spring, Fall/autumn or Rabi. Cultivars can be proffered based on length of growing season, availability of optimum moisture regime.

Sowing Time

Because of event of assorted climatic conditions in country, planting time differs from one spot to another. Ideal planting time in various agro-climatic districts is depicted in Table 2 [2]. The ideal opportunity to plant the yield relies upon accessibility of water system offices.

For instance, if water system offices are accessible, maize harvest can be planted around fourteen days before beginning of rainstorm while under rainfed conditions, crop is planted with the beginning of storm to have ideal dampness system so appropriate plant stand can be kept up in field.

Seed Rate

Being a non-tillage crop, it can't make up for the lost space if appropriate plant stand isn't kept up under field conditions. So, upkeep of 60–65,000 plants/ha is pre-imperative for acknowledging most extreme yield. Planting of the harvest ought to be done 60 × 20–25 cm crop math. For mixtures and composites, seed rate can be utilized as for seed weight and necessity of plant populace

Seed Treatment

Seed treatment assumes a crucial part in anticipation of infections and accessibility of supplements to developing yield. For example, seed treatment of maize with Bavistin or Derosal or Agrozim 50 WP (Carbendazim) @ 3 g/kg seed forestalls the assault of seed and soil borne sicknesses in maize crop. Moreover, treatment of seed with consortium (biofertilizer) @ 1.25 kg/ha helps in yield upgrade and improvement of soil wellbeing.

Crop Geometry

Harvest calculation has direct impact on entomb and intra-plant contest in field crops. Maize yield can be planted in changed harvest calculations (Table 5) relying on the motivation behind development [2, 8]. Inter-culture tasks like diminishing, hole topping and earthing-off assume basic part in execution of maize crop. Diminishing should be performed around 10 days after germination to keep 1 plant/slope. Further, 2 earthing-ups are needed in maize crop. First at 35–40 and second at 60–65 days after germination

Method of Planting

In spite of the fact that crop foundation is a progression of occasions that relies upon communications of seed, soil dampness, technique for planting, hardware and so on however strategy for planting assumes a significant part in foundation of harvest under given arrangement of conditions. Maize is essentially planted straightforwardly through seed by utilizing various strategies for culturing and foundation.

As of late, asset preservation innovations (RCTs) in particular, zero culturing, least culturing, surface cultivating and so forth had come by and by in different maize-based trimming framework and are savvy and climate amicable. Following are significant planting techniques that shift from one circumstance to another.

Zero tillage

Maize yield can be developed with no essential culturing under no-till with diminished expense of development and better asset use productivity. In the present circumstance, upkeep of legitimate soil dampness at planting and band arrangement of seed and manures with zero-till seed-cum-compost grower with wrinkle opener according to the dirt surface and field conditions is pre-essential.

The innovation is trailed by enormous number of ranchers particularly under rice-maize and maize-wheat frameworks in peninsular and eastern India. On the off chance that the field is plagued with weeds, ranchers can go for foliar shower of gram Oxone 24 SL (paraquat) @ 1250 ml/ha around 24 hours prior to planting of maize crop.

Ridge/Raised Bed Planting

This planting technique is viewed as best for development during rainstorm and winter seasons both under abundance and restricted water accessibility conditions. On non-uniform terrains, this technique is generally reasonable for fruitful development of maize crop. Planting of harvest should be done on the southern side of the east–west edges/beds for better openness to daylight during winters and better yield stand.

Raised bed grower having slanted plate, measuring or roller type seed dropping framework ought to be utilized for planting that works with legitimate position of seed and manures in single activity for having great yield stand, higher usefulness and asset use productivity.

Water system water can be saved to the tune of 20–30%. Under brief abundance soil dampness/water logging because of weighty downpours, the wrinkles will go about as seepage channels and yield can be saved from overabundance soil dampness stress.

Flat sowing

Maize yield can be developed by customary culturing level planting relying on soil type and accessibility of water system offices. Light soils have high penetration rate and low water holding limit, so ranchers can go for level planting of maize crop.

Under rain-fed conditions, to have better dampness accessibility to trim for longer period, level planting turns out to be better substitute. Level planting is additionally advantageous when no culturing framework gets plagued with high weed populace and compound/manual weed control becomes non-conservative.

Transplanting

It is better foundation procedure winter maize in the concentrated trimming framework where field can't be cleared on schedule, to forestall the postponed planting and harvest misfortune because of low temperature.

Under the present circumstance, nursery of the yield is raised on a more modest bit of land and seedlings are relocated in required field as and when they accomplish particular age. For instance, if the fields are to be emptied during December–January, it is prudent to go for nursery planting 30–40 days before the relocating. Seedlings can be relocated in the wrinkles followed by light water system.

Furrow planting

Furrow planting of maize is recommended when crop is to be cultivated during spring season as high evaporative losses may lead to water deficit stress in flat and raised bed or ridge sowing.

9.3.2. Water Management

Water prerequisite of the maize crop changes from 400 to 600 mm. Overabundance or lack of dampness can destructively affect the harvest development. Legitimate seepage of standing water and meeting the yield needs at basic stages assume a significant part in better harvest execution.

Particularly for winter maize, it is prudent to keep soil wet (incessant and gentle water system) during 15 December to 15 February to shield the harvest from ice injury.

Flood irrigation

Flood strategy for water system is followed where maize crop is developed with level planting. Yield is inundated as and when required. By and large, youthful seedlings, knee high stage (V8), blooming (VT) and grain

Furrow irrigation

At the point when yield is developed as edge/raised bed planting, furrow water system is followed. Care should be taken from the start water system that water ought not to flood on the edges/beds. As a thumb rule, the water system ought to be applied in furrows up to 2/third tallness of the edges/ beds.

In brought bed and up in restricted water system water, the water system water can likewise be applied in substitute furrow to save water system water. In rainfed conditions, tied-edges demonstrate supportive in saving the water, expanding its accessibility in the root zone for longer period.

Above ground drip irrigation

High temperature and high evaporative interest during summer season upgrades the water prerequisite of maize crop because of which ranchers go for various water system. To build the water use effectiveness of harvest, over the ground trickle water system is suggested by Punjab Agricultural University. In this, expansive beds are set up at 1.20 m separated from one focus to another of wrinkle. These beds are 80 cm wide on the top and 40 cm wide wrinkles between beds. The beds are covered with U.V settled plastic film (Black) of 25-micron thickness (23 grams for each m2). Two columns of maize are planted at a dividing of 60 cm keeping plant to plant distance of 20 cm. One horizontal line is utilized to flood two columns of maize.

The drippers are spaced 30 cm apart and are operated at a discharge of 2.2 L per hour. Prevailing climatic regimes of an area affect the efficiency of drip irrigation system.

Sub-Surface Drip Irrigation

In field tests, sub surface drip irrigation and fertigation brought about 18.4% higher framework efficiency with saving of 28.5% applied water system water. Sub-surface water system innovation can be continued in maize-wheat-summer moong trimming framework.

For this framework, Place trickle inline having dripper having 20 cm separating at 20 cm profundity with horizontal to parallel dispersing of 67.5 cm for sub surface dribble water system in maize-wheat-summer moong trimming framework.

Sow one line of maize, two lines of wheat and two lines of summer moong on each trickle inline during individual season. In the event that release of the dripper is 2.2 L/hour, the timetable given in Table 7 can be followed for sub-surface dribble water system in previously mentioned trimming framework

If discharge rate is different, then time of irrigation may be adjusted proportionally by the formula:

$$= (2.2 \times \text{time of irrigation (minutes)} \div \text{dripper discharge (litre/h)}$$

Partial root drying irrigation

This procedure includes substitute wetting and drying of two parts of root zone of harvest plants during back-to-back water systems. The PRD strategy was created based on information on root-to-shoot substance flagging (can be negative or certain) about soil conditions that manages the shoot physiology.

Exchanging is fundamental for keeping a consistent emanation of signs from the root-to-shoot, on the grounds that drawn out openness of root to drying soil may cause anatomical changes which decrease the capacity of root to detect soil drying and not ready to support the production of ABA for long time span.

Various strategies to apply the PRD strategy can be partition of root framework into two sections with sheet especially in pots, controlled substitute surface dribble water system on half piece of the root zone, controlled substitute subsurface trickle water system on half piece of the root zone or controlled substitute wrinkle water system.

9.3.3. Weed management

Maize crop is infested with grassy and broad leaf annual weeds. Among grassy, Dactylocteniumae gypticum, Eleucine indica, Setaria glauca, Cyanodondactylon, Cyperus rotundus, Sorghum helepanse, Brachariarapens are common. The broad leaf weeds are Celosia argentia, Commelina bengalensis, Phylanthisniruri, Solanum nigrum, Amaranthus viridis, Trianthemapartulacastrum.Compelling weed management procedures have key part in fruitful maize development.

Appropriation of weed control works on during the initial 6 two months subsequent to planting is pivotal in light of the fact that maize crop saved weed free for 30–45 days in the wake of planting is practically comparative in yield as that saved weed free for whole harvest period.

The yearly yield misfortune in maize due to weed issues is assessed to be roughly 10%. Various weed the executives drew closer can be followed for weed the board in maize crop that can be as per the following.

Non-Chemical Control: Manual Weeding, Mulching

Non-chemical weed control measures can physical or social that implies manual expulsion of weeds from the maize fields. In cultural method, given two hoeing 15–30 days after sowing with khurpa/kasaula/wheel-hoe/triphali/tractor-drawn cultivator. Mulching is practice of keeping crop residues or plastic sheets on the soil surface within the crop rows. Mulching helps in temperature regulation, water conservation as well weed control in field crops.

Chemical control

Sometimes due to continuous rains during the early stages of maize growth it becomes impossible to enter in the field. Also due to scarce availability of farm labour, the only effective way to control weeds is the use of herbicides.

Spray of atrataf 50 WP (atrazine) @ 2 kg/ha on medium to heavy textured soils and 1.25 kg/ha in light soils within 10 days of sowing, using 500 L of water prove propitious in keeping weed population low in maize fields.

Spray the herbicide uniformly at recommended rates to minimize residual toxicity to crops sown after maize. Alternatively, spray 262.5 ml/ha Laudis 420 SC (tembotrione) in 375 L of water at 20 days after sowing provides effective control of mixed weed flora. For the control of Cyperus rotundus (dila/motha), apply 500 ml/ha 2,4-D amine salt 58 SL as post emergence 20–25 days after sowing in 375 L of water.

9.3.4. Nutrient management

Figure 9.2. Nutrient management

Source: Image by agritech.tnau.ac.in

Integrated nutrient management

Among the oat crops, maize all in all and explicitly mixtures are receptive to supplements applied through natural or inorganic methods. The pace of use relies upon soil supplement status and editing framework. For acknowledging required yield, the portion of applied supplements ought to be as standard the dirt providing limit and harvest interest.

As the reaction of maize yield to natural fertilizers is noteworthy so coordinated supplement the board (INM) is vital alternative in maize-based frameworks. Apply 10–15 t/ha of good quality farmyard manure per hectare to the maize crop year after year.

Green fertilizer the field, to be put under maize with Dhaincha / Sunhemp / Cowpea. Cowpea / Dhaincha / Sun hemp ought to be planted during second fortnight of April utilizing 12/20/20 kg seed for each section of land, separately.

The 50 days old green fertilizer harvest ought to be buried and permitted to break down for around 10 days prior to planting of maize. On the off chance that, midyear moong crop is become the straw ought to be buried prior to planting of maize.

Inoculate the maize seed with suggested bio-compost as depicted prior. For this, blend half kg bundle of suggested consortium bio-manure with 1 L of water and afterward altogether blend it in with maize seed on clean pucca floor. Allow it to dry in shade and plant the seed right away. Vaccination with bio-compost ought to be done in the wake of treating the seed with fungicide. The seed vaccination with consortium bio-fertilizer increment grain yield just as improves soil wellbeing.

Paddy straw compost @ 450 kg/ha along with recommended dose of fertilizers can be an alternate to farm yard manure.

As a general recommendation, one could apply 120 kg N, 60 kg P2O5 and 40 kg K2O per hectare for hybrids and 80 kg N, 30 kg P2O5 and 20 kg K2O per hectare for composites. Drill 33% of nitrogen and the whole amount of phosphorous and potassium at the hour of planting.

Top dress 33% of nitrogen at the knee-high stage and the excess 33% at the pre adorning stage. It very well might be noticed that use of nitrogen compost more than suggested portion is not a viable alternative for FYM.

Diminished Zn accessibility visuals arise on center leaves (second or third from tip) of plants which incorporate white or light-yellow band and ruddy veins on the two sides of the midrib.

Fertigation

It alludes to synchronous utilization of water system water and manures by drip irrigation. By this strategy, FUE can go up to 80%. In dribble water system model for spring maize, certain suggestions are made in regard to manure application alongside trickle water system.

For the medium ripeness soils use of 200 kg of urea, 80 kg of mono ammonium phosphate (MAP) and 40 kg of muriate of potash (white)/ha is suggested. Start fertigation 12 days in the wake of planting of maize and apply 25% of the composts in four equivalent parts during first month on week-by-week premise. Rest of the manure ought to be applied in equivalent parts on week by week premise up to first seven day stretch of May. Besides, in sub-surface trickle water system, compost can be applied to maize crop when filled in maize-wheat-summer moong editing framework.

For example, Apply sub surface dribble water system at 3 days span for maize and summer moong with fertigation of 80% suggested portion of NPK. In maize, apply 1/5 dose of NPK at sowing and fertigate remaining P and K in 5 splits and N in 7 splits at 9 days interval starting from 15 DAS. Apply sub surface drip irrigation at 7 days interval up to mid-February and thereafter at 5 days interval to wheat with fertigation of 80% recommended dose of NPK. In wheat, apply 1/5th dose of NPK at sowing and fertigated the remaining NPK in 8 splits at 7 days interval starting from crown root initiation. In summer moong, fertigated NPK dose in 5 equal splits at 6 days interval starting from 10 DAS. Use urea, mono ammonium 119 phosphate and muriate of potash as source of N, P and K, respectively.

9.3.5. Insect and pest management

Integrated pest management (IPM)

IPM is exceptionally productive and eco-accommodating system which incorporates coordinated utilization of all potential substitutes that can be natural, physical, social or synthetic for controlling bugs. Producers who know about the potential for bug invasion follow a four-layered methodology. The four stages include: set activity edges, screen and recognize nuisances, avoidance and control.

Cultural control

Deep summer plowing helps in destroying resting stage of pests. Inter-cropping with legume reduces borer incidence. Use of well decomposed farm yard manure termite attack. Balanced use of fertilizers.

Genetic management

Use of good quality planting material from reliable source. Mechanical control: Cutting and destroying infected plants which ceases further spread.

Use of pheromone traps. Set up of light traps.

Chemical control

use of synthetic chemicals for the control of insect-pest and diseases.

Biological pest management

This approach encompasses use of living entities for the control of insect-pests and diseases. Living substances can be hunters, herbivores or parasites alongside concentrated human obstruction. For controlling maize drill and different creepy crawlies, apply bio-insect sprays like Neemazal (1%) @ 300 ml/ha.

The maize drill can likewise be overseen by utilizing tricho-cards twice having 40,000 eggs of Corcyra parasitized by Trichogrammachilonis.

Make first delivery on 10 days old yield and second multi week after first delivery. Cut tricho-cards into 40 equivalent strips and staple them consistently on the underside of the focal whorl leaves in evening hours. The tricho-cards ought not to be applied on blustery days.

Major insect-pests and diseases

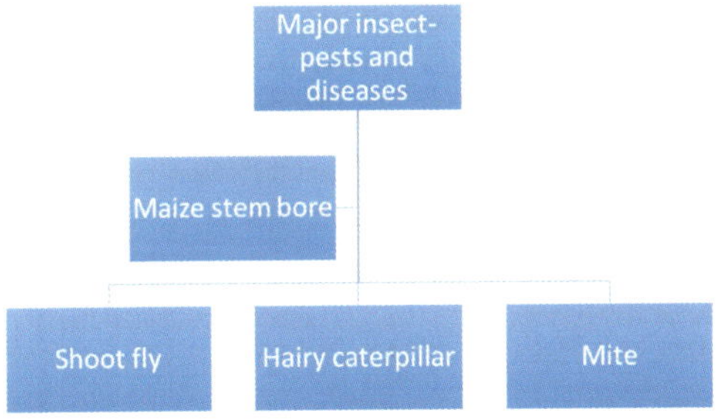

Maize stem bore

This insect attacks the maize crop principally developed during storm season. Subsequent to incubating, hatchlings enter the stem by scratching finished by exhausting whorl. Following procedures can be followed for avoidance and insurance of yield:

- Summer plowing of field.
- Destruction of perennating stages in stubbles, cobs, stalks.
- Cut and bury the severely infested plant parts.

Spray the crop 2–3 weeks after sowing as soon as borer injury to the leaves is noticed with Coragen 18.5 SC (chlorantraniliprole) @ 75 ml using 150 L water/ha with knap-sack sprayer.

Figure 9.3. Maize stem borer

Source: Image by SlideShare

Shoot fly

Although it is major pest in Southern India but it may infest the maize crop sown in spring season in Northern India. Mainly it attacks the seedling stage of crop where maggots move down to the basal portion through leaf sheath followed by cutting of growing point resulting in dead hearts. Control measures can be as follow:

- Spring crop should be sown between January 20 and February 15.
- Seed should be treated with gaucho (imidacloprid) 600 FS @ 6 ml/kg seed.

Hairy caterpillar

This pest becomes a serious concern when attacks in epidemic form. They damage the crop by feeding on leaves and soft stem, from gregarious (during younger stage) to distant migration (grown up stage). Prevention and protection strategies can be as follow:

- Collection and destruction of young larvae by cutting and burying the attacked plant parts.

- Physical destruction of large caterpillars.

Mite

The attack of mite is serious in June on the young crop or in September–October when the crop is nearing maturity. The affected leaves turn pale and can be recognized from the presence of dusty webs.

9.4. OTHER NEW TECHNOLOGIES AVAILABLE TODAY

Figure 9.4. Solar tech in maize production

Image by Farmers Weekly

Anyway, what else is accessible that could be portrayed as "cutting edge cultivating frameworks" or new advancements for the maize business? Lately, the selection of different exactness horticulture innovations by U.S. maize and soybean cultivators has been expanding drastically.

Among the more commonly embraced exactness cultivating advancements are those related with auto navigation of gear, auto-control of individual columns or spouts of use hardware, and variable rate control of use gear. Interest in information the board innovations is additionally developing as these abilities become more accessible and moderate.

The expanding accessibility and reasonableness of sub-inch (cm) exactness in GPS recipient advances has added to cultivators' advantage in auto-route and auto-control innovation lately. Specifically, selection of exact RTK-based GPS collector innovations is expanding quickly incompletely because of the always expanding accessibility of RTK networks all through the open country.

These organizations might be membership based (e.g., Trimble) or free access by means of state government branch of transportation RTK networks known as persistently working reference stations (CORS).

The benefits of auto-route and individual line control frameworks lie principally with their potential for diminishing expenses of yield production by uprightness of their capacity to decrease time and wastage related with over-and under-laps of field activities (culturing, compost spreading, herbicide applications, headlands or end rows).

Some potential for yield expands exists because of less skirts in weed control and less over-populace of headlands (end columns), however these yield upgrades will be moderately minor. I trust most would agree that none of the present exactness farming auto-nav or auto control innovations addresses a "quantum jump" in maize efficiency.

Maybe they offer the possibility to diminish production costs through more precise position of yield inputs. A theoretical, yet significant, advantage of auto-nav advances is the decrease in administrator weariness went with the opportunity to screen hardware all the more intently.

The chance for variable rate (VR) control for utilization of certain yield inputs, fundamentally manures and seeds, has been accessible for quite a while to U.S. agrarian retailers and progressively so at the farmer level lately. Manure retailers in the U.S. received VR application advancements for use of phosphorus and potassium composts in addition to lime as quite a while in the past as the 1990's.

Variable rate use of phosphorus and potassium manures in addition to lime has demonstrated compelling in decreasing entire field input costs and improving yields in spaces of fields with genuine lacks for P, K, or soil pH.

One reason this VR innovation has been compelling is that the agronomic information to help the VR choices can without much of a stretch be created from spatially escalated soil inspecting information that are then used to foster spatial application maps. All in all, the VR choices for these information sources depend on a sound comprehension of the agronomic connections between a solitary variable (spatial soil test information) and yield reaction.

While viable, these specific VR advances don't bring about "quantum jump" changes in the yearly pace of maize efficiency. Best case scenario, VR use of P, K, and lime improve the lower yielding spaces of fields recognized as lacking for these significant nourishing variables.

Variable rate control innovation is additionally accessible for nitrogen manure applications or cultivating rates. The two innovations are getting progressively accessible to ranchers and turning out to be normal standard extras on utensils and grower.

In the event that you make a "remedy" VR application record for use with a VR regulator, the instrument or grower will to be sure fluidly apply N rates or cultivating rates across a field dependent on what you advised it to do. In that lies the test. The product and equipment will do precisely what you advise it to do. Regardless of what premise you use to make an application map, the product will create a beautiful shading guide and "tell" the VR regulator what to do.

Rather than foreseeing P, K, and lime prerequisites dependent on soil test investigation information, there is no comparable single rule for N or cultivating rate necessities for maize with which to create solid VR application maps. At the end of the day, great deals of components interface to decide ideal N or cultivating rates for given spaces of a field.

Variable rate N application for a maize crop, in our experience, isn't identified with spatial yield potential. Maybe, VR nitrogen is, at its pith, identified with the assessed spatial changeability for soil N supply inside a given field. Notwithstanding, the gauge of accessible soil N supply is impacted by various elements that impact mineralization, immobilization, denitrification, draining, and volatilization of N.

These incorporate soil natural matter, inborn soil seepage attributes, ampleness of fake field waste, soil temperature, excrement application rates and techniques and timings, precipitation sums and timing and recurrence, timing of N manure application, strategy for N compost application, wellspring of N manure, and utilization of nitrification or urease inhibitors or not.

Making a VR nitrogen application map dependent on just a couple of these variables will frequently bring about flighty degrees of blunder in the genuine N prerequisites of the maize crop. The difficulties engaged with precisely anticipating soil N supply to thusly gauge variable N application rates have expanded the premium in utilizing optical reflectance sensors (e.g., Greenseeker™, Holland Scientific Crop Circle™) to appraise maize N status "in a hurry" and spatially side dress apply variable N rates as per the spatial fluctuation in crop shade reflectance.

Dynamic optical reflectance sensors are extremely delicate to changes in plant biomass and generally soundness of a maize crop. Different reflectance lists (e.g., NDVI, chlorophyll record) have been displayed to intently connect with N substance of plants or generally speaking N take-up by crops and furthermore connect intimately with possible grain yield.

The test with this innovation is that reflectance contrasts in plant N status because of variable accessible soil N are not apparent until the yield is excessively tall to side dress with conventional side dress line instruments.

The physiological justification this shortfall of reflectance contrasts in youthful maize plants is that maize basically doesn't start to take up critical measures of soil N until about leaf stage V9. Therefore, it is our assessment that optical harvest reflectance advancements are most appropriate for genuinely late side dress N application timings (~ V9 ~ 12) utilizing high-freedom tools (e.g., Hagie™, Miller™).

A late side dress timing procedure would probably incorporate a "close to planting" utilization of a base introductory N rate (maybe half - 75% of anticipated aggregate) trailed by a late side dressing application "controlled" by the optical reflectance information that would decide the amount of the excess 25% - half of the normal N was really required. All in all, the reflectance sensors would permit maize producers to "tweak" the use of the last 50+ kg of N compost.

Essentially to VR nitrogen rates, making variable rate cultivating rate maps is testing in light of the fact that there is no single model with which to base the "remedy". There is no single soil factor or other "the board zone" that obviously corresponds with ideal cultivating rates in maize. Ideal cultivating rates for maize are not connected as unequivocally with yield potential as one would envision.

In the U.S., the facts confirm that ideal cultivating rates are lower for testing conditions with perpetually low maize yields contrasted with less-testing conditions with higher maize yields. Actually, however, there are likely just two cultivating rates fundamental for some random field… … a higher rate and a lower rate. Apparently from past research that a surmised break point for lower versus higher ideal cultivating rates is beneath or over 8 T/ha. Throughout much of the U.S. Midwest, the lower optimal seeding rate for maize grain yield is about 62,000 seeds per hectare (sph), while the higher optimal seeding rate is no more than about 82,000 sph. My own field-scale research in Indiana has been led principally on ranches with yield levels more prominent than 8 T/ha, however unmistakably shows that there is no connection between [optimal cultivating rate] and [yield at ideal cultivating rate] for yield levels between around 8 and almost 14 T/ha.

The reach in ideal cultivating rates for maize grain yield in 27 field-scale preliminaries lately has been fundamentally between 61,000 to 87,000 sph with a normal ideal cultivating pace of 74,000 sph.

9.4.1. Data management technologies

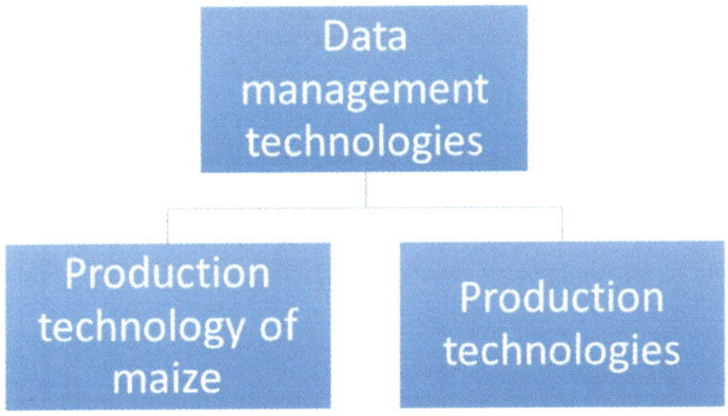

Figure 9.5. Data management technologies

True modifications and advancements in cab display screens, regulators, handheld gadgets, and remote correspondence innovations are enormously improving our abilities to plan, screen, log, and send rural information from work vehicle to office to field to other gear to office to landowner to inhabitant to project worker to specialist like nothing we have seen previously.

This upgraded capacity to oversee information electronically and spatially is really magnificent. Such devices improve our capacity to distinguish spatially reliable regions in fields that merit our consideration and in a real sense permit us to explore to those spaces for determination and remedial measures. Such devices may improve our capacity to oversee hardware efficiencies and lower our expenses of creation.

9.5. PRODUCTION TECHNOLOGY OF MAIZE

Spread overseas in mid-1500's. Early sorts were stones and flours. Extensive inconstancy was because of open pollination. Dent types emerged about 1500 years ago. 1940's: Hybrid corn seed. The six significant kinds of corn are dent corn, rock corn, unit corn, popcorn, flour corn, and sweet corn.

The verdant tail of the plant produces separate dust and ovuliferous inflorescences or ears, which are organic products, yielding parts or seeds. Maize bits are frequently utilized in cooking as a starch. Maize is a famous fall/autumn grain in Pakistan and discovers a lot of utilization as grub for animals and poultry.

The yield can be reaped in 60 – 65 days and it very well may be filled in many pieces of the country. The green tail has high starch substance so it's anything but a captivating feed for poultry. Mixture seed ought not to be utilized for grub purposes since they have less foliage, and short height with less sustenance.

Hybrid seed corn improvement Develop stable 'innate' lines. Cross two distinctive ingrained lines. Sow the seed delivered (F1 cross breed). Posterity will yield 2-5X guardians.

New advancements are herbicide safe (glyphosate and some others) and inner insect spray creating transgenic assortments. New advancements are herbicide safe (glyphosate and some others) and inward insect spray creating transgenic assortments.

One year crop rotation-

- Maize – wheat
- Maize – Barseem
- Maize – B.T Cotton

Two-year crop rotation-

- Maize – wheat – maize – barseem
- Maize – wheat – cotton – barseem

Soil and climate- Best growth can be found on fertile, well drained, medium-textured soils. Salinity and water logging are harmful at seedling stage. Suitable temperature for germination is 21-32 C. Better yield of maize is observed at pH 5.5-8 of soil.

Continuous water logging for 3 days reduces the yield by 40-45%. Temperature >35 C reduce the pollen. Temperature < 15 C delay the silking and tasseling. 500-750 mm rainfall is required for growth.

9.5.1. Production technologies

Land preparation- Two to three furrowing in the wake of collecting. Cost of hal 500-600 for furrowing. Use rotavator to blend the dirt and to debase the coagulations 1200-1500. Use ridger to make edges and wrinkles and the expense of the ridger per section of land is 1000rupees.

Keep up the distance between wrinkles roughly 75cm. Line to push distance 26-28inches. Plant to plant distance ought to be 6inches (cross breed) for early planting. 8-10 inches (half breed) for late planting Plowing has significantly diminished as of late as planting gear opened up that permit seedling in to standing yields.

Conservation tillage system prevents soil disintegration, save water, and diminishes cost of creation. Corn can be utilized with all protection culturing framework. Edge planting Flat planting The land with ideal dampness after precipitation or water system is extricated 20-25 cm profound through furrowing or disking 2-3 times, quickly followed by planking. Planting techniques Ridge planting

Manual strategy (chopa technique); its normal in Pakistan. Tractor drawn ridger; ridges are made by ridger and the distance between the two ridges 27inches. P×P= 20-25 cm R×R= 75 cm Flat sowing 1. Automatic tractor drawn drill or manual/hand drill R×R= 75 cm.

Thinning after 10-15 days of emergence Plant population 70,000 to 75,000/ha sowing times Jan-march for early sowing called as spring crop and maturity of crop within 50-60days. And yield of crop comes in 100-120 Munds. Jun-Aug. for late sowing called as autumn crop in this maturity of crop increases than 60 days. And yield comes from it is 60-80 Munds 1. Spring 1 week of February – 1 week of March. 2.

Autumn Last week of July – mid of August. Seed rate 10-12 kg/acre or 35000 seeds/ acre Seed treatment immediately after the emergence of plant from Seed to prevent from shoot fly and other insects. Seed is treated with fungicides before sowing, so it may prevent from these. Telsa & Aghaz is used as fungicides.

9.6. CONCLUSION

This chapter gives insight to the reader by addressing the production technologies which are new and improved and are applied in the production of maize in order to enhance and make the production high.

Due to the use of all these advanced technologies the production of maize is day by day increasing. The chapter starts with elucidating the crop establishment, the selection of cultivar, sowing time, seed rate, seed treatment, crop geometry.

It sheds light on the different methods of planting, the zero tillage, ridge or raised bed planting, flat sowing, as well as transplanting. It then describes the water management which further deals with flood irrigation, furrow irrigation, above ground drip irrigation, sub-surface drip irrigation and the partial root drying irrigation.

As the chapter proceeds, it talks about the management of weed which incorporates various dimensions such as the non-chemical control which is manual weeding and mulching, it then highlights nutrient management like the integrated nutrient management together with explaining the fertigation.

It also addresses the insect and pest management and biological pest management processes. Towards the end, it puts emphasis on the different types of enhanced technologies which are available in the current year and will be available in the future.

The various advanced technologies which is present in the recent years is the data management technologies, production technologies, and the manual strategies.

REFERENCES

1. AnsaRebi,2019.Productiontechnologyofmaize.[online]ResearchGate. Available at: <https://www.researchgate.net/publication/334131857_Production_technology_of_maize> [Accessed 30 June 2021].

2. Jaidka, M., Bathla, S. and Kaur, R., 2020. Improved Technologies for Higher Maize Production. Maize - Production and Use, [online] Available at: <https://www.intechopen.com/books/maize-production-and-use/improved-technologies-for-higher-maize-production> [Accessed 30 June 2021].

Chapter 10

Challenges in Maize Production

CONTENTS

10.1. Introduction 220

10.2. Problems and Issues for White Maize Production 226

10.3. Difficulties and Opportunities in Maize Production 228

10.4. Fertilization and Liming 230

10.5. Crop Rotation 232

10.6. Development and Cultivation 232

10.7. Problem Descriptions 232

10.8. Management Factors 236

10.9. Maize in the Developing World: Trends, Challenges, and Opportunities 240

10.10. Conclusion 242

References 244

This chapter highlights the different types of challenges and difficulties prevailing in the production of maize. It begins with addressing the problems and issues for white maize production. Then it further talks about the difficulties and opportunities in the production of maize.

As the chapter proceeds it showcases the concept of fertilization and liming. It then deals with the factor of crop rotation along with the development. It also puts emphasis on the problem descriptions by depicting the different aspects which affect the cultivation such as the climatic factors and soil conditions, mineral nutrition and chemical injury etc.

10.1. INTRODUCTION

Maize is the reason for food security in a portion of the world's least fortunate districts in Africa, Asia, and Latin America. Along with rice and wheat, maize gives in any event 30% of the food calories of more than 4.5 billion individuals in 94 non-industrial nations.

They incorporate 900 million customers with livelihoods of not exactly US$ 2 every day for which maize is the favored staple. Universally, 765 million metric huge loads of maize were reaped in 2010 from only less than 153 million hectares.

Around 73% of this space was situated in the creating scene, with again an overwhelming extent of this space in the low and lower center pay nations. Maize is presently created on almost 100 million hectares in 125 agricultural nations and is among the three most broadly developed yields in 75 of those nations.

The yield gives more than 20% of complete calories in human eating regimens in 21 nations, and more than 30% in 12 nations that are home to a sum of in excess of 310 million individuals. Production of maize, particularly in the tropical areas, is influenced by various limitations, including a variety of abiotic and biotic anxieties, helpless soil richness, absence of admittance to key information sources (particularly quality seed and manures), low degrees of automation, and helpless post-gather the board.

The outcome: maize yields in a considerable lot of the sub-Saharan African nations, where maize is the main staple food, are frequently very low, averaging roughly 1.5 tons per hectare—about 20% of the normal yield in created nations — and yields in a few Asian nations are still under 3 tons for each hectare.

Figure 10.1. Challenges in maize production

Source: Image by unsplash.com

Furthermore, 33% of all malnourished children are found in frameworks where maize is among the main three harvests. Repetitive food value emergencies—joined with the worldwide monetary emergency, unpredictable energy costs, regular asset exhaustion, and environmental change—undermine the jobs and food security of millions of destitute individuals. Simultaneously, interest for maize in the creating scene is relied upon to twofold by 2050.

By 2025, maize is expected to have become the yield with the best production worldwide and in the developing world. The growth in demand for human consumption of maize in the developing world is predicted to be 1.3% per annum until 2020.

Also, rising wages are relied upon to bring about a multiplying of utilization of meat across the creating scene, prompting an anticipated development popular for feed maize of 2.9% per annum. Be that as it may, harvests at current degrees of usefulness development will in any case miss the mark regarding request and a great many homestead families will stay in neediness. Unless vigorous measures are taken the result will be more expensive nourishment for a great many helpless maize buyers, proceeding with neediness and youth hunger, deforestation, soil debasement, decreased biodiversity, and sped up exhaustion of water and compost saves. Creating and sending environment strong germplasm adjusted to the tropical/sub-tropical maize-developing locales has gotten one of the highest needs. A similarly significant undertaking is to upgrade the reception of such germplasm by the smallholder ranchers.

While reproducing and conveyance of value seed are without a doubt basic for the eventual fate of maize creation, and therefore, the vocations of a few million smallholder ranchers around the world, these endeavors alone won't be satisfactory.

Supportable heightening of maize-based frameworks warrants complementation of improved cultivars with maintainable harvest and regular asset the board rehearses, just as financial mediations for maize fates (successful arrangements, establishments, innovation focusing on, and markets).

"MAIZE – Global Alliance for Improving Food Security and the Livelihoods of the Resource-Poor in the Developing World" is the CGIAR Research Program (CRP) that aims to address the above-mentioned challenges of enhancing global food security, reducing poverty, and sustaining the environment in the maize-based systems.

MAIZE, fundamentally, presents another results-oriented technique to completely misuse the capability of global maize research-for-improvement. Other than rearing elite maize germplasm with tolerence to different abiotic stresses (particularly dry spell, heat, waterlogging, and causticity), another important challenge while adapting yields to the changing environment will be to keep up their hereditary protection from insect-pests and diseases.

Despite the fact that methodologies to restrict the impact of environmental change on irritations and sicknesses don't on a very basic level vary from existing incorporated pest management practices, there should be a lot more prominent accentuation on displaying and gauging frameworks, while have opposition will keep on having an essential job.

Ongoing examination has prompted the improvement of a set-up of soil and yield the executives practices for expanding asset use productivity while keeping up soil wellbeing, and moderating ozone depleting substance discharges. Expanded utilization of protection farming practices is needed to diminish environmental change impacts.

As well as recognizing/fostering another age of maize cultivars appropriate for protection agribusiness-based works on, accomplishing expanded transformation activity will require mix of environmental change-related issues with other danger factors, like environment fluctuation and market hazard, and with other arrangement areas.

Maize gives food, feed and dietary security in a portion of the world's most unfortunate areas in Africa, Asia, and Latin America. Internationally, 765 million metric tons (m t) of maize were gathered in 2010 from just shy of 153 million hectares (m ha).

Around 73% of this space was situated in the creating scene, with again a prevalent extent of this space in the low and lower center pay nations. The harvest gives more than 20% of absolute calories in human weight control plans in 21 nations, and more than 30% in 12 nations that are home to an aggregate of in excess of 310 million individuals (Shiferaw et al. 2011).

For 900 million ranchers and shoppers in low-and center pay nations, maize is a favored yield or food. The development, popular for human utilization of maize in the creating scene, is anticipated to be 1.3% per annum until 2020.

Also, rising wages are relied upon to bring about a multiplying of utilization of meat across the creating scene (Naylor et al. 2005), prompting an anticipated development sought after for feed maize of 2.9% per annum.

In any case, the normal maize yields in a few of the agricultural nations, particularly in sub-Saharan Africa, where maize is a profoundly significant staple food crop, are still under 1 t/ha, while numerous nations have just 1-2 t/ha, due mostly to helpless soil fruitfulness, continuous event of dry seasons, high frequency of creepy crawly vermin, sicknesses and weeds, ranchers' restricted admittance to manure, and absence of admittance to improved maize seed (Shiferaw et al. 2011).

The significance of improving maize production and usefulness in the creating scene could be checked from the way that 33% of all malnourished kids are found in frameworks where maize is among the best three yields (Hyman et al. 2008) Asia's commitment to the overall gathered maize region just as production has been essentially expanding.

The significant maize makers in Asia are China and D.P.R. Korea in East Asia; Indonesia, Philippines, Vietnam and Thailand in Southeast Asia; India, Pakistan and Nepal in South Asia; and Turkey and Iran in West Asia.

It is remarkable that eight significant maize-delivering nations in Asia – China, India, Indonesia, Nepal, Pakistan, Philippines, Thailand, and Vietnam – taken together, presently produce 98% of Asia's maize and 26% of worldwide maize (Erenstein, 2010); in this load of nations, maize is transcendently become under rainfed conditions by the smallholder, asset helpless ranchers.

The maize situation in Asia is fairly special contrasted with the remainder of the world. Firstly, 70% of the total maize produced in Asia is used for feed purposes, 23% as food, and 7% for other uses. By contrast, in sub-Saharan Africa, maize is mainly a food crop accounting for 73% and 64% of the total demand in Eastern and Southern Africa (ESA) and Western and Central Africa (WCA), respectively (Shiferaw et al., 2011).

Despite the fact that the maize feed market is quickly developing, particularly in nations like China, India and Indonesia, maize is as yet a significant staple food in numerous nations/regions in Asia, particularly in the slopes and ancestral areas of Nepal, Bhutan, and India.

Besides, as far as grain inclination, not at all like sub-Saharan Africa where white maize assumes an exceptionally predominant part as food, in practically all the Asian maize-developing nations, the interest is generally for yellow maize.

Regardless of these distinctions, the asset helpless maize ranchers in Asia face numerous difficulties that are shared by smallholders in sub-Saharan Africa and Latin America; among many, these incorporate helpless buying limit, a variety of abiotic and biotic burdens, helpless soil ripeness, and restricted admittance to quality seed (especially in the non-business maize belts).

The world is encountering rising interest for harvests like maize. It has been assessed that worldwide horticultural production should increment by 60% or more by 2050. A few projections propose an especially popularity for maize over other cereal harvests.

To satisfy this expanding need will need around a 2.4% each year expansion in yield. Nonetheless, in view of current execution, gauges propose yearly worldwide yield increments for maize as low as 1.6%, well underneath the rate required.

This normal likewise hides critical varieties between districts. While maize yield expansions in Asia are by and large high, in South America they change somewhere in the range of 1.7 and 4%. Varieties in maize yield increments are generally sensational in Africa where a few nations are confronting diminishes in yearly yields of more than 7%. As a now yields in sub-Saharan Africa can be just about as low as 1.5 tons per hectare, a fifth of what is accomplished by the main maize makers.

Projections by the International Food Research Policy Institute (IFRI), using its International Model for Policy Analysis of Agricultural Commodities and Trade (IMPACT), suggest cereal yields in Sub-Saharan Africa may still be less than a third of the highest yields in East Asia by 2020.

A new Foresight Report by the UK government proposes a hole among potential and genuine maize yields of more than 60% for Sub-Saharan Africa. It is broadly acknowledged that expanding production should be accomplished by improving yields as opposed to clearing more land for development.

Indeed, even in those spaces where all reasonable land isn't as of now under development, there is expanding rivalry for land, water, work and different assets (for example for urbanization), also developing worry about the effect of deforestation and land leeway on soil quality and richness (for example because of issues like disintegration, fermentation and saltiness), environmental change and biodiversity.

There is consequently a need to improve yields by making horticulture more effective, beneficial and maintainable, especially for smallholders in those locales, for example, sub-Saharan Africa which join a high dependence on maize for food with low degrees of usefulness and more weak agrarian frameworks. Right now, maize development in these nations is kept down by various issues.

One bunch of difficulties identifies with abiotic stresses. Most maize development in the creating scene is reliant upon precipitation. This makes the harvest especially defenseless against dry season and warmth. Current assessments recommend that 25% of maize production is compromised by regular dry spell.

It is generally acknowledged that these anxieties will be gotten more intense because of environmental change, bringing about an expected loss of up to 10% in maize production in Africa and Latin America by 2055. Other abiotic stresses incorporate poor/corrupted soil quality which is likewise seen as a huge contributing component to helpless yields.

Maize cultivation is also affected by a range of biotic stresses including:

- Diseases such as downy mildew, rust, leaf blight, maize streak virus (MSV) and more recently, maize lethal necrosis (MLN) which has led to the total loss of maize crops in some parts of Kenya

- Insect pests, for example, the stem borer which can causes misfortunes of 20-40%

Other anxieties, for example, the parasitic Striga (witch) weed which has prompted misfortunes of 60-90% in certain pieces of Nigeria Finally maize development is kept down by an absence of assets and preparing for smallholders.

These requirements incorporate absence of admittance to improved assortments and great quality seed, restricted admittance to frequently costly manures and pesticides, helpless spread of good agrarian practice, the requirement for reasonable, limited scope automation to supplant physical work and critical post-reap misfortunes because of helpless stockpiling, bug assault and decay.

10.2. PROBLEMS AND ISSUES FOR WHITE MAIZE PRODUCTION

Extending white maize yield will rely basically upon expansions in yields, which thus will expect ranchers to embrace new innovation. From an ecological perspective, in numerous nations expansions in maize yields on better land are desirable over extending maize development into more negligible regions.

The utilization of improved seed-manure innovation stays sketchy in many non-industrial nations developing white maize, as does the utilization of more refined, the executives escalated advances. For new advancements to be accessible to ranchers, dynamic agrarian examination programs that consider the requirements of customer ranchers, just as augmentation benefits that work effectively with ranchers as they find out about new strategies for maize development, are important.

In any case, all throughout the planet exploration and expansion spending plans are contracting in genuine terms at both the public and worldwide levels.

A portion of the issues with the take-up of improved maize assortments and half and halves have come about because of deficient enthusiasm for farmer inclinations and conditions in the advancement of these materials.

Around the world, be that as it may, the best imperative to the dispersion of better germplasm to ranchers has been the disappointment of seed ventures to advance to address the issues of a wide assortment of cultivators. Discussion has frequently fixated on the overall propriety of public and private area interest in innovative work, production and showcasing, and the serious part of general society/private area relationship has been focused.

More creative methodologies are vital. These ought to perceive the probability that public and private areas can be reciprocal, that their jobs may advance after some time, and that different sorts of entertainers, for example, farmer gatherings and non-administrative associations, may offer helpful types of assistance.

Figure 10.2. Problems and issues for white maize production

Source: Image by unsplash.com

Inventive methods of extending the extent of harvest the board research are likewise required, with the goal that the expenses of creating both efficiency improving and asset monitoring innovations are decreased. Across the maize-developing creating world, the administration of soils and soil richness is the most squeezing crop the board issue as of now.

The information base concerning the support of soil natural matter in the jungles, under smallholder conditions should be significantly extended. Also, both rural and strategy research coordinated at more proficient utilization of inorganic manures is important. Especially in sub-Saharan Africa, the need to make manure utilize more effective is dire on account of high compost costs.

In numerous nations, a large number of varieties in field size and strategies for development exist among various gatherings of ranchers. Rural examination and strategy should resolve the topic of the general accentuation to be given to the tiny, transitional or enormous farmer gatherings.

This inquiry is muddled by the way that both normal field size and disparity of asset conveyance shift broadly from one country to another, so approaches must be adjusted to the particular arrangement of monetary and social boundaries portraying a country.

The responses to these issues are probably going to be convoluted; however the significance of white maize as a bland staple warrants deliberate endeavors to address them. Without complementarity of strategy change and innovation advancement, white maize production in many agricultural nations is probably going to fall further behind request development, making them more reliant upon a worldwide market described by a serious level of instability.

10.3. DIFFICULTIES AND OPPORTUNITIES IN MAIZE PRODUCTION

Maize is the main grain crop in Southern Africa and costs are drastically affected by any sign of under or over creation. The harvest is similarly as significant in different pieces of the world, and a lot of it is created if climatic conditions are positive. This prompts a worldwide over-creation, which impacts contrarily on the maize cost.

A valid example is the thing that occurred with the South African maize value last year because of the dry season which influenced enormous pieces of the maize creating regions. There was a deficiency of maize and this caused ruin. Luckily, enough maize could be imported to make up the deficiencies and forestall the purported "yellow porridge" dread.

The maize cost is dictated by organic market. Both yield and grain costs are drivers in the rancher's business. Promoting of grain consistently sounds simple on paper, yet it's anything but significantly more confounded by and by. There can likewise be a considerable expense included, contingent upon the showcasing channel utilized.

Accordingly, the maker's test is still to get the most ideal costs for his produce, and to create however much grain as could be expected with the assets available to him. This sounds actually quite difficult, yet on the off chance that we are not receptive to transform, it will be hard to accomplish.

Expansions in grain costs or yields won't generally occur in enormous augmentations. Nonetheless, it's difficult the huge enhancements that matter, however the little, steady changes in crop costs and yield likewise have an effect. Any expansions in grain cost or yield ought to never be to the detriment of long-haul supportability.

Figure 10.3. Difficulties and opportunities in maize production

Source: Image by unsplash.com

Because of new hereditary qualities, maize yield increments are simpler to accomplish, gave that none of the components adding to yield are restricted in any capacity. This is exceptionally occasional the situation practically speaking.

Ecological conditions, for example, soil dampness, convenient precipitation, waterlogged soils, supplement inadequacies, weed pressure, plant infection and irritations are regularly reasons why the harvest's hereditary potential isn't reached. Indeed, even in a more controlled climate, for example under water system, it's anything but consistently conceivable to arrive at the hereditary potential.

A maize yield of 4 tons per hectare (t/ha) at a price of R2 250/ton will earn a profit of 28.6% (at a production cost of R7 000/ha). If the maize price increases with a mere R250 per ton (R2 500/ton), the profitability will increase by 14.3%. If the price remains the same, but the yield increases with 1 t/ha with an increased production cost of R1 000/ha more, therefore R 8 000/ha, profitability will improve by 12%. It does not necessarily cost

R1 000 per hectare more to produce an extra ton of grain. If the producer succeeds in achieving a ton per hectare yield increase, as well as a R250 per ton better maize price, the total profitability can improve by 26.3%.

The grain market (Safex) does offer options to put maize price management into place. Use the opportunities offered by the market, but do this with the advice of a good grain marketer.

Having the option to oversee yields is the ideal for some ranchers, since the executives practices impacts it straightforwardly. The worth of good and right counsel is frequently thought little of. Yet, the worth of guidance that isn't restricted to simply treatment however which centers on every single agronomic perspective, generally surpasses the worth of the compost costs per hectare. The Omnia agronomist is the foundation of this interaction. Agronomic methodologies depend on presumptions, yet on innovative work led and affirmed over the long haul.

Without innovative work, the pace of progress will decay quickly and this could prompt a circumstance where more food should be imported, which will have adverse results for the South African economy. To guarantee that we don't get behind the remainder of the world, Omnia will proceed to create and deliver a-list items. The essential focal point of our innovative work is to diminish the rancher's danger and to guarantee that maintainability isn't abandoned.

Which agronomic viewpoints can be addressed to guarantee that maize arrives at its full ecological potential?

10.4. FERTILIZATION AND LIMING

Compost can be utilized with care to maneuver a plant toward a specific course. Perhaps the main parts of plant nourishment are to guarantee that the right compost item is utilized. The right situation of the manure is basic to guarantee that plant roots will actually want to utilize the supplements and convert them into grain.

The measure of compost applied needs to praise the capability of the dirt and climate. Don't finished or under treat, as both are impeding. Treat as per the potential and apply the fundamental supplements at the ideal opportunity with the goal that the plant can use it productively.

Tending to the above angles in regards to treatment ought to improve the supplement use productivity. This implies that the sum and sort of manure is utilized ideally and the greatest amount of grain per unit supplement created.

Figure 10.4. Fertilization and liming

Source: Image by unsplash.com

On the off chance that the right sort of manure is utilized, the water use effectiveness ought to likewise improve. The plant will create more grain per millimeter of water under dry-land conditions, just as under water system.

Expanding sub-soil acidification is a worry, as it very well may be inconvenient to the drawn-out productivity of maize, particularly on sandy soils, except if it is remedied. Ideal root development and advancement can be limited at a pH (KCl) of lower than 5.5.

A low soil pH is likewise connected with a high corrosive immersion rate (aluminum poisonousness), however regularly additionally with a low calcium and magnesium supply. In the particular model the pH (KCl) levels were changed to 5.8 in the dirt (0-200 mm) and 4.7 in the sub-soil (300-500 ml) one year after the right sort and amounts of lime were applied.

Sub-soil acidity is only sometimes remedied after a brief period and it must be observed routinely. Nitrogen manure must be applied with care as band setting huge amounts could advance sub-soil acridity, contingent upon the nitrogen source. Decline the pre-plant application and attempt to apply a bit of the nitrogen compost as a topdressing. This will likewise improve nitrogen use productivity.

10.5. CROP ROTATION

Crop rotation is certainly a significant part of effective grain cultivating. It is, nonetheless, not generally conceivable as the harvests which could be utilized in pivot are not generally fit to the dirt kind or development framework.

To understand the additional huge load of yield alluded to before in this article, crop rotation is one of the fundamental practices to consider. Pick the rotation crop cautiously to guarantee a positive response on the ensuing maize yield. Harvest pivot can likewise be gainful in lessening sickness or weed pressure.

10.6. DEVELOPMENT AND CULTIVATION

Soil ought to be developed barely enough to guarantee most extreme yields for a specific season. Soils change, and an activity that is fruitful in one spot won't really chip away at the following ranch. In this manner, first look at the dirt and afterward decide how it ought to be developed.

Development ought to basically advance dampness protection and great root improvement, yet without presenting the dirt to erosion, for instance. To ensure that the soil on your farm reaches its full potential, please contact your nearest Omnia agronomist.

10.7. PROBLEM DESCRIPTIONS

10.7.1. Climatic Factors and Soil Conditions

Moisture stress

Water is the most widely recognized impediment to maize production in the jungles. Dry season during the harvest foundation stage can kill youthful plants, diminishing the plant thickness. The fundamental impact of dry season in the vegetative period is to decrease leaf development, so the yield catches less daylight.

Around blossoming (from around fourteen days prior to silking to about fourteen days in the wake of silking), maize is touchy to dampness stress. Grain yield can be truly influenced if dry spell happens during this period.

Figure 10.5. Climatic factors and soil conditions

Source: Image by unsplash.com

During the grain-filling period, the primary impact of dry spell is to lessen part size. By and large, maize needs in any event 500-700 mm of well distributed precipitation during the developing season. Indeed, even that measure of downpour may not be sufficient, in any case, if the dampness can't be put away in the dirt on account of overflow or shallow soil profundity, or if the evaporative interest is huge because of high temperatures and low relative moistness.

The event of dampness stress as a rule differs incredibly from one year to another. On the off chance that you notice indications of stress one year, you actually need to analyze climate records and converse with ranchers to see whether it's anything but a typical issue. On the off chance that dry spell is a typical issue, it will decrease yield by over 20% in one year out of four.

Water logging

The maize crop is quite sensitive to water logging, that is, to saturate soils. From planting to the V6 stage, water logging for more than 24 hours can kill the crop (especially if temperatures are high) because the growing point is below the soil surface at that time. Later in the crop cycle, water logging can be tolerated for periods of up to one week, but yield will be reduced significantly.

Sun and Temperature Impacts

The maize harvest can endure a wide scope of temperatures (from 5 to 45oC), however exceptionally low or high temperatures can negatively affect yield. Generally, a farmer can't do a lot to change the impact of temperature other than shift the planting date marginally or plant a superior adjusted or before developing assortment.

Maize varieties do vary fundamentally in their temperature reactions. The harvest is normally not harmed by high daylight except if it is likewise under temperature or dry spell pressure. The harvest is frequently influenced by low daylight when broadened times of overcast climate happen, particularly in the event that they harmonize with blossoming.

Once more, the farmer regularly can't do a lot to change the measure of daylight accessible to the yield, yet it is significant that you can perceive the manifestations related with these issues so you don't mistake them for different elements.

Lodging

Maize is regularly influenced by Lodging, both of the root or tail. A plant is supposed to be root stopped if the lower tail frames a point of 45o or less with the dirt surface. It is tail stopped if the tail is broken beneath the ear and the wrecked segment frames a point of 45o or less with the dirt.

Root Lodging and tail dwelling are regularly inadequately connected; root dwelling will in general be related with ecological factors, for example, weighty downpours corresponding with wind or the board factors like high densities or helpless plant circulation, while tail breakage is frequently firmly identified with hereditary attributes like illness and creepy crawly obstruction, productivity, and senescence designs.

The impact of Lodging on yield relies upon when it happens and whether the ears kept in touch with the dirt long enough for spoiling or germination to happen. Financial misfortune from Lodging additionally relies upon the collecting strategy utilized in the locale. In the event that collect is by machine, many held up plants won't be reaped. In the event that the farmer reaps by hand, housing will expand the time required and work costs.

Figure 10.6. Sun and temperature impacts on maize fields

Source: Image by unsplash.com

Acid or alkaline soils

The maize crop by and large develops above and beyond a pH scope of 5.5-7.8. The impact of soil pH outside this reach is as a rule to make certain components pretty much accessible, so toxicity or inadequacy creates.

Under a dirt pH of 5.5, Al and Mn poison levels are oftentimes issues, and insufficiencies of P and Mg are normal. At a pH over 8 (or over 7 on calcareous soils1), lacks of Fe, Mn, and Zn will in general happen. The field side effects of unacceptable pH accordingly for the most part take after those of micronutrient issues. Saline soils- Maize is considered sensitive to salinity.

When the electrical conductivity of a saturated soil extract is 2.5 milli siemens/cm (mS/cm, same as millimhos/cm), you can expect a 10% yield reduction; a value of 4 is associated with a 25% yield reduction. This yield loss usually results from the fact that the plants cannot extract enough water from the salt-affected soil. In some cases, the salts are actually toxic to the crop.

10.8. MANAGEMENT FACTORS

Planting issues: Land planning and planting strategies there are three kinds of issues which happen at planting. One is that the land may not be developed as expected, so that there are lumps or outside layers which forestall planting at a uniform profundity and even germination.

Another is that planning is excessively well before planting or is inconsistent, so the weeds enjoy an upper hand over the harvest. The third is that the seeds are set at some unacceptable profundity. The main roles of land planning before planting are to make a dirt construction great for crop development, to join buildups, and to control weeds and sicknesses.

In numerous spaces, the dirt construction is satisfactory to permit great development without development, insofar as weeds are constrained by different strategies. Deposits can regularly be left on the dirt surface, on the off chance that they don't make other field activities troublesome.

A decent planting strategy is one that permits seed to be set at the right profundity and gives great contact among seed and soil. The right profundity of planting is sufficiently profound to permit seed to take up water, to shield it from drying up or birds, and to keep it from growing with light rains, yet shallow enough to permit the seedling to arrive at the surface prior to exhausting its food saves or being assaulted by soil creepy crawlies or sicknesses.

The reasonable planting profundity for swamp tropical maize is normally around 5-7 cm, yet however profound as 10 cm may be frequently adequate, if the seed is enormous and solid. When planting to utilize lingering dampness, particularly in high country regions, farmers may place seed at profundities of 20 cm.

Be that as it may, particular assortments are required for planting at such profundities if the seed is covered totally. At the point when seed is planted in dry, exposed soil in warm regions, the profundity ought to associate with 10 cm to stay away from high temperature harm.

Thickness The ideal thickness under non-restricting conditions will be distinctive for various assortments, and it ought to be resolved for the significant assortments in your locale. The optimum density could be related roughly to plant height and maturity for CIMMYT lowland tropical germplasm growing in one environment.

The ideal thickness at reap for an assortment is what yields the most grain when the harvest is become under non limiting conditions, which are never found in farmers' fields. Hence, the suggested thickness at collects (what will give the best grain yields in ranchers' fields) will be not quite the same as the ideal thickness.

The best thickness for grain yield in a farmers' field will shift from one year to another, contingent upon the environment and harvest the executives. An agronomist working in a space needs to discover the populace thickness which will give the farmer the best return in terrible years just as in great years.

It has been shown that a decrease in populace thickness of 30% beneath the ideal will just lessen yields by about 5% in great years, and the lower thickness should expand yields when stress happens. Accordingly suggested densities are for the most part around 20-30% beneath the ideal thickness.

In case, the dry season pressure is extremely normal in your locale, the suggested densities will require to be changed descending significantly over 30% from the ideal to build the water accessible to each plant. In semi-parched districts, the danger of harvest disappointment increments forcefully as thickness builds.

In the event that the thickness is diminished to adapt to dry season, recall that weed control may likewise require more consideration, since the low thickness maize will give less shade. Whenever you have settled on a suggested thickness, you should ascertain the measure of misfortune you anticipate from planting to reap to get a suggested cultivating rate.

Look at the reap thickness nearby with the quantity of seeds the farmer plants. In numerous conditions, the deficiency of plants from planting to collect is around 20%. Gap your suggested thickness by one less the % misfortune (1 - % misfortune) to get the suggested cultivating rate.

For example, if a material is being grown with an optimum density of 85,000 plants/ ha, the recommended density for farmers' fields might be: 85,000 - (85,000 x 0.30) = 60,000 plants/ha However, it would be expected about 20% of the plants to be lost between planting and harvest due to insect attack and disease. The recommended seeding rate is thus: 60,000 ÷ (1 - 0.20) = 75,000 seeds/ha If there are 3,500 seeds in a kilo, that rate equals 21.4 kg of seed per hectare.

Plant distribution- Since hand planted maize is regularly organized in slopes (singular planting stations) with more than one plant for every slope, spatial course of action can at times be an issue, even where thickness is proper. Studies show that when multiple plants fill in a similar slope, grain yield is influenced by rivalry for water, supplements, and light. In the event that at least four plants fill in a similar slope, one-to-three of them as a rule don't create an ear. For what reason do farmers regularly sow numerous seeds in a single slope? At the point when the farmer plants by hand, making more seed openings takes longer.

Just in case, the farmer makes more openings per hectare, it should be made sure that the yield benefit will be beneficial. Moreover, farmers in some cases overplant and afterward eliminate plants during the season as scrounge for animals.

Mineral nutrition- The maize crop needs certain mineral components in sufficient amounts for great development. These supplements are for the most part provided from the dirt and from added manures. Albeit the maize plant utilizes 13 unique supplements, just three are needed in moderately enormous amounts: nitrogen, phosphorus, and potassium.

These are the supplements which most much of as far as possible maize creation, however sulfur and a few micronutrients, for example, zinc and magnesium might be significant imperatives in certain spaces.

Yield can often be reduced 10-30% by deficiencies of major nutrients before any clear symptoms of deficiency are observed in the field. Even if you don't see symptoms, it is a good idea to calculate a nitrogen balance and analyze soil P levels.

Chemical injury-Maize can be harmed by the inappropriate utilization of farming synthetic compounds like herbicides, composts, or insect sprays. The harm is normally the aftereffect of applying the compound indiscreetly, at too high a rate, at some unacceptable development stage, or when the plants are under dry spell or temperature stress.

Generally, issues of synthetic injury are the aftereffect of incidental mishaps and are not significant constraints to yield in a space, however it is significant that an agronomist have the option to perceive these issues. Herbicide injury can cause contortion of youthful plants and yellowing, consuming, and kicking the bucket of the leaves.

The deformity, (for example, contorted leaves which don't unroll as expected, or twisted roots) comes about because of applying herbicides from specific gatherings, for example, 2,4-D at some unacceptable development stage or at too high a rate.

This problem can occur with phenoxy herbicides (e.g., 2,4-D), dinitroaniline herbicides (e.g., pendimethalin), benzoic acid herbicides (e.g., dicamba), and amide herbicides (e.g., alachlor or metolachlor). Leaf chlorosis and death can result from over application of triazine herbicides (e.g., atrazine).

Leaf burning comes about because of getting a directed spray of a contact herbicide like paraquat on the maize or from applying too high a convergence of another substance. You can recognize herbicide injury from foliar infection by searching for copying which happens in designs from the sprayer spout and which just influences leaves of a specific age which were uncovered when the compound was applied.

If composts (particularly N and K) are put in touch with the seed or excessively near youthful plants they can cause copying, particularly if soil dampness is inadequate. On the off chance that the issue happens at planting, the seeds may not grow, or the seedling may arise and afterward bite the dust. Consumed plants may look shriveled or hindered.

At the point when manure is side dressed, it ought to be set no nearer than 10 cm to the stem. Foliar applied insect poisons or composts can periodically cause consuming of the maize leaves. This can be limited by applying promptly in the first part of the day or late in the evening, to stay away from extreme, direct daylight on the leaves.

Weed Competition - The significance of weed rivalry in maize relies upon four factors: the harvest development stage, the measure of weeds present, the level of water and supplement pressure, and the weed species. Weeds harm the harvest basically by viewing for light, water, and supplements.

Maize is sensitive to this opposition during the basic time frame between the V3 and the V8 stages. Before the V3 stage, weeds are normally significant just on the off chance that they are bigger than the maize or if the harvest is experiencing water pressure.

Maize needs a period between the V3 and V8 stages when few weeds are present. From the V8 stage to maturity, the crop usually reduces the sunlight reaching the weeds enough to provide good weed control. In the later piece of the cycle, weeds are significant primarily if water or supplement pressure is an issue, or if a forceful weed overtops the maize and shades it, or if the weed has some allelopathic impact.

Furthermore, a few weeds make reaping troublesome, and along these lines increment production costs. Some weed species because more harm than others. This can be on the grounds that the weeds really produce harmful substances which harm the harvest (allelopathy) or in light of the fact that the weeds are extremely compelling contenders for water or supplements

10.9. MAIZE IN THE DEVELOPING WORLD: TRENDS, CHALLENGES, AND OPPORTUNITIES

Regardless of the great development somewhat recently, the normal maize yields in a considerable lot of the Asian nations actually stay low opposite the world normal of 5.02 t/ha in 2010.Countries with sizable maize area but with less than 3 t/ha yields include India, Indonesia, Philippines, Pakistan and Nepal, while Vietnam and Thailand have registered maize yields in the range of 4-5 t/ha in 2010. China recorded an average yield of 5.33 t/ha in 2010.

In spite of the fact that maize is filled in pretty much every territory in China, around 66% of the maize in China is filled in calm, high-potential production conditions in the north, and the rest is filled in the subtropical and tropical conditions of the south. Dry season is perceived as the main requirement across the rainfed marsh and upland conditions, covering about 70% of the maize production region in Asia.

The present circumstance is probably going to worsen in the coming a very long time because of environmental change, frequently prompting insufficient and additionally lopsided occurrence of precipitation in the yield season close by temperature changes (IPCC 2007).

Lightening the impacts of dry season alone could build normal maize yields by 35% across Asia-7 (barring China), and by 28% in Southwest China (Gerpacio and Pingali 2007). Simultaneously, more than 18% of the complete maize production region in South and Southeast Asia is regularly influenced by floods and waterlogging issues, causing production misfortunes of 25–30 % yearly (Zaidi et al. 2010).

Before this present century's over, developing season temperatures will surpass the most outrageous occasional temperatures recorded in the previous century (Battisti and Naylor 2009). Recent analysis of more than 20,000 historical maize trial yields in Africa over an eight-year period, combined with weather data, showed that for every degree day above 30°C, maize grain yield was reduced by 1% and 1.7 % under optimal rainfed and drought conditions, respectively (Lobell et al. 2011).

High temperature stress and drought are probably going to disturb in northern China (Piao et al. 2010) just as in numerous tropical maize developing regions, particularly in South and Southeast Asia. Spring maize is a significant alternative for heightening and differentiating trimming frameworks in South Asia, yet is inclined to extreme warmth stress during blossoming/

early grain filling stages, especially in the upper and center Indo-Gangetic fields.

This features the significance of creating maize germplasm with resistance to both dry season and high temperature stress. Biotic burdens that that have far reaching impacts in Asia incorporate the wool molds, post-blooming tail spoils (PFSR), dark leaf spot (GLS), joined leaf and sheath curse (BLSB), turcicum leaf scourge (TLB), ear decays, mycotoxins, stem drills and weevils.

Rising temperatures and variations in dampness additionally possibly influence the variety and responsiveness of microbes and bug bothers, and could prompt new and maybe eccentric epidemiologies (Gregory et al. 2009).

For example, GLS is now becoming an important disease globally, with high incidences reported in Nepal, China, Bhutan, Colombia, Mexico, Brazil and several countries in Africa. Poor soil fertility (including micronutrient deficiencies) and low nutrient use efficiency also rank among the most important factors limiting crop productivity and yield stability in both high potential–low risk environments as well as low potential–high risk environments.

Figure 10.7. High temperature stress and drought probably disturbs the maize yielding

Source: Image by unsplash.com

A common concern is imbalanced fertilizer use (i.e., very high use of N, less use of P, and negligible use of K, S, and micronutrients), which is particularly prevalent in the rice maize systems that are distributed all over South Asia, but more so in Bangladesh, India, Nepal, and Pakistan (Timsina et al. 2010).

Clearly, hereditary qualities and reproducing alone can't tackle the intricate test of improving efficiency in smallholder ranches. There is an unmistakable requirement for powerful complementation of improved maize cultivars by appropriate preservation agribusiness rehearses.

Accessibility of hardware for direct cultivating or insignificant culturing tasks is a requirement that should be tended to in growing locally applicable choices for protection farming. Moreover, agronomists and geneticists/reproducers need to work couple to recognize cultivars that react best to such practices, and for producing better comprehension of the mind-boggling associations between genotype x climate x administration rehearses.

Much must be accomplished likewise as far as institutional and strategy advancements that help maize development and improvement in Asia. Likewise, an incorporated methodology that connects the biophysical and the financial work is fundamental for accomplishment in improving efficiency of maize and in upgrading variation to evolving environment (Shiferaw et al. 2011).

This incorporates understanding the smallholder ranchers' moderateness and admittance to quality seed, limitations in reception of high-yielding, stress tough and healthfully improved maize assortments, and organizations and arrangements to altogether upgrade seed production and circulation.

10.10. CONCLUSION

This chapter gives an insight to the reader by portraying the various issues, challenges and difficulties faced in the production of maize which needs to be overcome and tackled with. It begins with providing a detailed account of problems and issues for white maize production. It then illustrates the prevailing difficulties as well as the opportunities in maize production.

It gives a brief idea into the fertilization and liming by describing them. As the chapter proceeds it mentions the concept of crop rotation, the development process etc. Towards the end, it highlights the problem descriptions by addressing the various factors which affects the production such as the climatic factors and soil conditions, mineral nutrition, chemical injury etc. It also puts emphasis upon the weed competition and then in the end, illustrates in detail the production of maize in light of the developing world and the trends, challenges as well as the opportunities it provides to people in different countries.

REFERENCES

1. BM Prasanna, n.d. Maize in the Developing World: Trends, Challenges, and Opportunities. [ebook] Nairobi, Kenya: Proceeding International Maize Conference, pp.26 - 38. Available at: <https://pangan.litbang.pertanian.go.id/files/IMC-PDF/03-Prasanna.pdf> [Accessed 30 June 2021].
2. FAO, n.d. X. PROBLEMS AND ISSUES FOR WHITE MAIZE PRODUCTION. [online] Fao.org. Available at: <http://www.fao.org/3/w2698e/w2698e0a.htm> [Accessed 30 June 2021].
3. H. R. Lafitte, n.d. IDENTIFYING PRODUCTION PROBLEMS IN TROPICAL MAIZE A FIELD GUIDE. [ebook] CIMMYT. Available at: <https://afghanag.ucdavis.edu/grain-field-crops/files/maize-cimmyt.pdf> [Accessed 30 June 2021].
4. Kobus van Zyl, 2017. Challenges and opportunities in maize production - Omnia Nutriology®. [online] Fertilizer.co.za. Available at: <http://www.fertilizer.co.za/knowledge-centre/expertise/229-challenges-and-opportunities-in-maize-production> [Accessed 30 June 2021].
5. n.d. Maize: key challenges. [ebook] Burleigh Dodds Science Publishing. Available at: <https://bdspublishing.com/_webedit/uploaded-files/All%20Files/Maize%20blog%20key%20challenges.pdf> [Accessed 30 June 2021].

INDEX

A

accreditation body (AB) 148
agriculture 54, 65, 67, 71, 77
agroclimatology 54
Agro-Ecosystem Analysis (AESA) 166
animals feed 2, 12, 23

B

Bacillus thuringiensis (Bt.) 3
biofuel 2, 4
biological control 166, 169, 189
biological pest management 192, 216
bio pesticides 166, 189

C

Cellulosic ethanol 4
certificate body (CB) 148
chemical injury 220, 243
Clayey soils 33
climate 54, 56, 58, 62, 63, 65, 74, 75
climatic factors 220, 242
cloddier seedbeds 33
consumer products 2, 22
Corn 2, 4, 5, 15, 20, 24
corn ethanol 4
corn flour 4
corn plant 2, 5
Crop agriculture 81
crop growth 54, 61, 69
crop inspection 80
Crop rotation 232
crop weed competition 122
crop yields 80
cultural practices 166, 189

D

Detasseling 97
dirt 28, 29, 31, 37, 38, 39, 45, 47, 48, 50
drip irrigation 192, 201, 202, 205, 206, 216
drying 80, 96, 98
dry matter (DM) 146

E

Eastern and Southern Africa (ESA) 224
Eastern Gangetic Plains (EGP) 123
Economic Threshold Level (ETL) 166
eight amaranth 131
ethnobotany 2

F

farmers' practices (FP) 146
fertigation 192, 202, 206, 216
fertilization 102, 107, 118
flowering 102, 103
Food Security 222
fungicides 166, 181, 189

G

genetic management 166, 171
good agricultural practices (GAP) 146, 164
grain 83, 85, 87, 91, 97, 99

H

helicoverpa armigera 166, 189
herbicide glyphosate 3
herbicides 122, 123, 125, 132, 133, 135, 138, 139, 140, 141
high soil fertility (HSF) 146
horticultural creation 126
human food 2, 22
humidity 54, 55
hybrid seed production 80, 87, 95

I

insecticides 166, 189
insect pests 166, 168, 170, 189
Integrated agronomic practice (IAP) 146
integrated crop management (ICM) 146
integrated nutrient management 192, 216
integrated pest management (IPM) 146
International Food Research Policy Institute (IFRI) 225
International Institute for Tropical Agriculture (IITA) 30

K

kernel numbers 102

L

Land 27, 28, 29, 41, 46, 47, 49, 50, 51, 52
Land quality pointers 29
landscape frames 28
leaf area index (LAI) 146
light 54, 55, 56, 60, 61, 70, 75
light quality 102
low soil fertility (LSF) 146

M

maize 2, 3, 5, 6, 7, 8, 9, 10, 11, 12, 13, 14, 15, 16, 17, 18, 19, 20, 21, 22, 23, 24
Maize cultivation 225
maize lethal necrosis (MLN) 225
maize production 28, 46, 51
maize streak virus (MSV) 225
manual weeding 192, 216
meat 221, 223
mechanical practices 166, 189
mineral nutrition 220, 242
Mulching 34, 35

N

national seed regulations 89
natural sources 122
nitrogen use efficiency (NUE) 146
nutrient management 192, 197, 204, 216

O

organic fertilizers 30
organic interaction 54
organic matter 34, 42

P

Pest monitoring 166
Pests 166, 170, 172, 186, 187
phosphorous 30
photoperiod 102
plantation 28
planting, 192, 196, 199, 201, 216
planting conditions 102
pollen sheds 102
pollen tubes 102, 116, 118
potassium 30

R

rainfall 54, 60
raw material 2
risk assessment approach 146
root 102, 105, 108, 109, 110, 111, 112

S

Scheme Owner (SO) 148
Seed 79, 80, 81, 82, 84, 85, 86, 89, 90, 91, 93, 94, 95, 98, 100
seed germination 102, 108
shelling 80, 86, 90, 98
silking 102, 115, 116, 117, 118
soil conditions 220, 233, 242
soil fertility 54
soils 54, 60, 61, 63, 64, 65, 66, 67, 68, 69, 71, 72, 74
Stalks 5
stem 103, 104, 105, 110, 112
sunshine 54, 60
surface hydrology 28
sustainable agriculture 146

T

temperature 102, 107, 108, 111, 115, 117
tillage 192, 196, 198, 199, 215, 216
transplanting 192, 216

V

variable rate (VR) 210
Vegetables 58
vegetation 30, 35, 46

W

water 54, 55, 56, 57, 60, 61, 62, 64, 67, 68, 69, 70, 71, 72, 73, 74, 75, 76
Water management 96
weed control 122, 123, 125, 130, 132, 133, 134, 137, 138, 139, 140, 141, 142
weed management 80
Weeds 122, 123, 125, 126, 127, 128, 142, 143
Western and Central Africa (WCA) 224
wind 54, 60, 70